and
Romano-British Antiquities

1/7/2002

From Leaders to Rulers

FUNDAMENTAL ISSUES IN ARCHAEOLOGY

Series Editors: Gary M. Feinman
The Field Museum, Chicago, Illinois
T. Douglas Price
University of Wisconsin-Madison

Editorial Board: Ofer Bar-Yosef, *Harvard University* • Christine Hastorf, *University of California-Berkeley* • Jeffrey Hantman, *University of Virginia* • Patty Jo Watson, *Washington University* • Linda Manzanilla, *Universidad Nacional Autónoma de México* • John Parkington, *University of Capetown* • Klavs Randsborg, *University of Copenhagen* • Olga Soffer • *University of Illinois* • Matthew Spriggs, *Australian National University* • John Yellen, *National Science Foundation*

EMERGENCE AND CHANGE IN EARLY URBAN SOCIETIES
Edited by Linda Manzanilla

FOUNDATIONS OF SOCIAL INEQUALITY
Edited by T. Douglas Price and Gary M. Feinman

FROM LEADERS TO RULERS
Edited by Jonathan Haas

LANDSCAPES OF POWER, LANDSCAPES OF CONFLICT
State Formation in the South Scandinavian Iron Age
Tina L. Thurston

LEADERSHIP STRATEGIES, ECONOMIC ACTIVITY, AND INTERREGIONAL INTERACTION
Social Complexity in Northeast China
Gideon Shelach

LIFE IN NEOLITHIC FARMING COMMUNITIES
Social Organization, Identity, and Differentiation
Edited by Ian Kuijt

A Continuation Order Plan is available for this series. A continuation order will bring delivery of each new volume immediately upon publication. Volumes are billed only upon actual shipment. For further information please contact the publisher.

From Leaders to Rulers

Edited by

JONATHAN HAAS
The Field Museum
Chicago, Illinois

Kluwer Academic / Plenum Publishers
New York   Boston   Dordrecht   London   Moscow

Library of Congress Cataloging-in-Publication Data

From leaders to rulers/edited by Jonathan Haas.
    p.   cm. — (Fundamental issues in archaeology)
Includes bibliographical references and index.
ISBN 0-306-46421-7
    1. Political anthropology—Congresses. 2. Social evolution—Congresses. 3. Social archaeology—Congresses. I. Haas, Jonathan, 1949–. II. Series.

GN492 .F76 2000
306.2—dc21

                                                                                                                      00-062212

ISBN: 0-306-46421-7

©2001 Kluwer Academic / Plenum Publishers
233 Spring Street, New York, New York 10013

http://www.wkap.nl/

10 9 8 7 6 5 4 3 2 1

A C.I.P. record for this book is available from the Library of Congress

All rights reserved

No part of this book may be reproduced, stored in a retrieval system, or transmitted in any form or by any means, electronic, mechanical, photocopying, microfilming, recording, or otherwise, without written permission from the Publisher.

Printed in the United States of America

Preface

The evolution of complex cultural systems is marked by a number of broad, sweeping patterns that characterize many different cultures at different points in time across the globe. Over the course of the past 100,000 years, there has been a general evolutionary trend for cultural systems to get larger and more complex. A consistent element in the broad course of cultural evolution has been the emergence and subsequent development of centralized forms of political organization.

The record of the first modern humans illuminates a global wide pattern of relative social equality and decentralized decision-making processes. Prior to about 10,000 years ago, there are no indications of clear social, political, or economic hierarchies. In these early millennia archaeological markers of social ranking are lacking and there is a similar absence of evidence pointing to the presence of leaders, chiefs or rulers. The pattern of social equality began to change at different moments and at different rates in various parts of the world in the course of the last 10,000 years. In some areas, such as Mesopotamia, politically centralized hierarchies emerged very early and developed rapidly, while in others, such as the Arctic, political centralization never emerged outside the context of Western colonialism.

In every culture area, the origins and development of politically centralized social systems and the emergence of leaders and rulers followed a unique evolutionary trajectory depending on local history and environment. At the same time, there are broad patterns of similarity in the evolution of politically centralized polities that cut across cultures and across time. This volume considers both the patterns of similarities as well as significant differences in the evolution of political centralization in diverse environments across the globe. The authors explore how and why leaders emerge out of egalitarian foundations, how chiefs evolve into state rulers, and how rulers expand their realms into empires. They examine how rulers gain and enhance their bases of power and how the role of individual rulers may or may not impact the evolution of entire cultural systems.

The collection stems from a workshop and conference held at The Field Museum in Chicago in October 1997. The idea of the conference first grew out of ongoing discussions between Chip Stanish, then chair of the Anthropology Department at the Museum, and myself about how and why political systems evolve. We felt that bringing together a group of scholars working in different world areas would be productive in a couple of ways.

First, it would counter the trend toward the inevitable geographical provincialism that prevails in archaeology. People working in a particular area tend to communicate and interact mostly with people working in that same area; in doing so they are not always able to take full advantage of alternative perspectives and data sets of researchers working in other areas. Bringing together scholars from different areas allows for an intellectual cross-fertilization and for a broader familiarization with cultural sequences in other landscapes.

Second, it would provide an opportunity to integrate analyses of centralization across a wide spectrum of complex political systems. Again, there is all too often a tendency for archaeologists working on certain kinds of societies, that is, states or chiefdoms, to look only at those societies and not fully consider antecedent and descendant political processes. As in the case of geographical diversity, our goal was to promote an intradisciplinary synergy and arrive at a better, broader understanding of how and why political systems evolve.

The conference participants came together for a two-day closed workshop in which the individual papers were discussed. Abbreviated versions of the papers were then presented to a public audience in an all-day conference. Although he was the person who had the original idea for the conference, Chip Stanish was ultimately unable to attend; we missed his good humor, his perspective from the Andean highlands, and his thoughtful insights. Anne Underhill participated in the workshop/conference but was not able to contribute to the volume; her views, growing out of her work on emergent complex societies in China, were very welcome. On behalf of all the participants, I would like to express our appreciation to The Field Museum for providing the support of the conference. I would also like to thank Dan Corkill for his patience and persistence in compiling the bibliographies, formatting the papers, tracking down illustrations, and the myriad other tasks so unpleasant, but integral to the editing process.

Contributors

Brian R. Billman • Department of Anthropology, University of North Carolina-Chapel Hill, Chapel Hill, North Carolina 27599

Winifred Creamer • Department of Anthropology, University of Northern Illinois, DeKalb, Illinois 60115

Carole L. Crumley • Department of Anthropology, University of North Carolina-Chapel Hill, Chapel Hill, North Carolina 27599

Timothy Earle • Department of Anthropology, Northwestern University, Evanston, Illinois 60208

Gary M. Feinman • Department of Anthropology, Field Museum, Chicago, Illinois 60605

Antonio Gilman • Department of Anthropology, California State University-Northridge, Northridge, California 91330

Jonathan Haas • Department of Anthropology, Field Museum, Chicago, Illinois 60605

Kristian Kristiansen • Institute of Archaeology, Goteburg University, Goteborg, Sweden

Patricia A. McAnany • Department of Anthropology, Boston University, Boston, Massachusetts 02215

Gil Stein • Department of Anthropology, Northwestern University, Evanston, Illinois 60208

Contents

PART 1. INTRODUCTION

Chapter 1 • Cultural Evolution and Political Centralization 3

Jonathan Haas

Introduction 3
Savagery, Barbarism, and Civilization........................... 4
Bands, Tribes, Chiefdoms, and States............................ 6
Selection and Transformation 10
Tinkering and Trajectories..................................... 13

**Chapter 2 • Communication, Holism, and
the Evolution of Sociopolitical Complexity 19**

Carole L. Crumley

Introduction 19
Evolution, Self-Organization, and Chaos 20
Organizational Complexity: Hierarchy and Heterarchy 24
Authority Structures: Hierarchies and Heterarchies 25
    Hierarchical Polities....................................... 26
    Heterarchical Polities...................................... 26
    Trade-offs 27
Heterarchies in Business: Matrix Organization 27
Liminality 28
An Application: State Hierarchy and Environmental Change 29
The State of the State 31

PART II. THE EMERGENCE OF LEADERS

Chapter 3 • The Origins of Centralization: Changing Features of Local and Regional Control during the Rio Grande Classic Period, ad 1325–1540 37

Winifred Creamer

Demographic Change	38
Economic Change	41
Thirteenth Century	42
Fourteenth Century	42
Fifteenth Century	42
Aggregation	48
Warfare	51
Ritual and Religion	55
Conclusions	56

Chapter 4 • Assessing Political Development in Copper and Bronze Age Southeast Spain 59

Antonio Gilman

Introduction	59
Environmental Background	61
Archaeological Sequence	61
Traditional Interpretation	64
Processual Alternative	65
Metallurgy	68
Agricultural Intensification	70
Management or Exploitation?	73
Rulers or Leaders?	77
Conclusion	80

PART III. LEADERS TO RULERS

Chapter 5 • Rulers and Warriors: Symbolic Transmission and Social Transformation in Bronze Age Europe 85

Kristian Kristiansen

Symbolic Structures and Social Institutions	85

CONTENTS

The Twin Rulers: The Ritualized Structure of Political Leadership 88
    Tracing the Meaning of a Symbol: The Chiefs Cap 88
    Tracing the Contexts of the Twin: Dualism 91
    Origins of the Twin Rulers 96
The Social and Cultural Context of Warrior Aristocracies 98
Symbolic Transmission and Social Transformation
    in Bronze Age Europe 103

Chapter 6 • Institutionalization of Chiefdoms Why Landscapes Are Built 105

Timothy K. Earle

Introduction 105
Background Literature 108
Three Archaeological Cases: Denmark, Peru, and Hawaii. 111
    The Cultural Landscape of Thy, Denmark in the Late Neolithic
        and Early Bronze Ages (2400–1200 BC) 113
    The Cultural Landscape of the Mantaro Valley, Peru
        in the Wanka II Period (AD 1300–1460) 116
    The Hawaiian Cultural Landscape during the Formation
        of Regional Chiefdoms (AD 1200–1800) 120
Conclusions 124

Chapter 7 • Cosmology and the Institutionalization of Hierarchy in the Maya Region 125

Patricia A. McAnany

Power, Politics, and Time 126
Formative to Classic Period in the Maya Lowlands 128
Kin Hierarchies and Individual Status Roles 130
Interactive Polities that Mutually Reinforce Each Other 134
Warriors and Military Engagements 136
Narratives of Heroism 138
Writing as a Means of Kinship Validation and a Key to Governance .. 139
Exclusive Access to Ancestors and Supernaturals through
    Shamanistic Ritual 141
Wealth and Its Aura 144
Final Thoughts on the Rise of Rulership in the Maya Lowlands 147

PART IV. RULERS IN POWER

Chapter 8 • Mesoamerican Political Complexity
The Corporate–Network Dimension 151

Gary M. Feinman

Introduction 151
Monolithic Notions of Hierarchy 153
The Corporate and Network Modes 155
The Classic Maya and Teotihuacan 161
Reflections on Mesoamerican Calendrics 168
The Corporate Mode: Comparative Utility 170
Concluding Thoughts 173

Chapter 9 • Understanding the Timing and Tempo of the Evolution of Political Centralization on the Central Andean Coastline and Beyond 177

Brian R. Billman

The Process of Centralized Political Formation 179
The Central Andean Coastline 186
The Late Preceramic Period 188
The Initial Period 191
Leader–Follower Relationships in the Late Preceramic and
    Initial Periods 195
Explaining the Timing and Tempo of Pristine Centralized
    Polities on the Central Andean Coast 199

Chapter 10 • "Who Was King? Who Was Not King?" Social Group Composition and Competition in Early Mesopotamian State Societies 205

Gil Stein

Introduction 205
Changing Perspectives on the Organization of Mesopotamian States .. 211
    Widespread Urbanism 211
    Endemic Warfare 211
    Emergence of Kingship 211
    Pronounced Social Stratification 211
    Increasing Use of Writing 212

Earlier Models of Centralized Mesopotamian States 212
States as Heterogeneous Social Networks 214
"Sectors" in Early Mesopotamian State Societies 217
    The Palace 217
    The Temple 217
    Large Estates 218
    Craft Specialists 218
    Urban Commoners 218
    "Semifree" Individuals 219
    Slaves 219
    Villagers 219
    Nomads 219
The Archaeology of "Invisible" Sectors in Mesopotamian
    State Societies 221
Craft Production and "Dual Economies" in Mesopotamia 222
"Resilient" Village Production in an Urbanized Regional System 226
Conclusions 231

PART V. CONCLUSION

Chapter 11 • Nonlinear Paths of Political Centralization **235**
Jonathan Haas

References **245**

Index **283**

Part I

Introduction

Chapter **1**

Cultural Evolution and Political Centralization

JONATHAN HAAS

INTRODUCTION

Since the inception of the discipline, the study of cultural evolution has played a cyclical role in anthropology. Over the past 125 years, anthropologists have tended to either emphasize or deemphasize the importance of long-term evolutionary change in the course of human history. The first direct analyses of cultural evolution emerged in the latter half of the nineteenth century as social scientists in Europe and the United States began to struggle with the diverse people and cultures of Africa, Asia, the Americas, and other world areas. The field of anthropology emerged as a social science umbrella devoted to the politics, sociology, economics, art, and ideology of these non-Western peoples from the far corners of the globe. Lewis Henry Morgan and Edward B. Tylor were among the first anthropologists to begin writing synthetically about non-Western peoples. Both Morgan and Tylor developed formal evolutionary models as a means of bringing order to the plethora of information being brought in from the field by missionaries, colonists, and nascent ethnographers.

JONATHAN HAAS • Department of Anthropology, The Field Museum, Chicago, Illinois 60605.

From Leaders to Rulers, edited by Jonathan Haas. Kluwer Academic/Plenum Publishers, New York, 2001.

SAVAGERY, BARBARISM, AND CIVILIZATION

Working relatively independently, Morgan (1877) and Tylor (1871, 1881) identified broad patterns of similarity in many different cultures around the world and they each proposed evolutionary typologies for organizing these crosscultural patterns. Both Morgan and Tylor divided cultures into three basic stages: savagery, barbarism, and civilization. Axiomatic to both models was the principle that all societies at the "civilization" stage had gone through the other two stages and those at the "savagery" or "barbarian" stages if given enough time would evolve into "civilizations." Both scholars characterized the three stages by an explicit list of shared attributes. Tylor looked at different aspects of culture, such as language, mythology, the "arts of life," or the "arts of pleasure." In his discussion of each, he considered the traits and conditions that prevailed under the different stages and how later characteristics evolved out of earlier ones:

> Modern music is thus plainly derived from ancient. But there has arisen in it a great new development. The music of the ancients scarcely went beyond melody.... The musical instruments of the present day may all be traced back to rude and early forms. The rattle and the drum are serious instruments among savages; the rattle has come down to a child's toy with us, but the drum holds its own in peace and war. Above these monotonous instruments comes the trumpet, which, as has just been seen, brings barbaric music a long step further on. In the modern [civilized] orchestra, the cornet is a trumpet provided with stops. (Tylor 1881:164–165)

Morgan takes a somewhat complementary perspective by looking at broad evolutionary patterns including the growth of intelligence, growth of the idea of government, growth of the idea of the family, and growth of the idea of property. In his analysis of government, for example, he traces the development from organization of society on the basis of sex characteristic of savagery as found among the Australian Aborigines, through the barbarian kin-based organization of the Iroquois and other North American Indians, to the hallmark of civilized society, the institution of Roman political society. There is a pervasive sense in the work of both Tylor and Morgan that cultural evolution represents "progress" in some external objective sense:

> The educated world of Europe and America practically settles a standard by simply placing its own nations at one end of the social series and savage tribes at the other, arranging the rest of mankind between these limits according as they correspond more closely to savage or to cultured life. (Tylor 1871:23)

> The latest investigations respecting the early condition of the human race are tending to the conclusion that mankind commenced their career at the bottom of the scale and worked their way up from savagery to civilization through the slow accumulations of experimental knowledge.
> As it is undeniable that portions of the human family have existed in a state

of savagery, other portions in a state of barbarism, and still other portions in a state of civilization, it seems equally so that these three distinct conditions are connected with each other in a natural as well as necessary sequence of progress. (Morgan 1877:3)

As a subsequent generation of anthropologists gained new and better information about cultural diversity, the progressive stage models of Morgan and Tylor were criticized and largely rejected in the opening decades of the twentieth century. The particularist school of Franz Boas and his colleagues focused on the unique individuality of cultures and turned away from analyses of general patterns of cross-cultural similarities and evolutionary change. There was a sense that culture was not a phenomenon that could or should be explained through scientific principles. The primary task of anthropology was to describe the historical path and inner working of individual cultures (Boas 1911). The comprehensive evolutionary stages of Morgan and Tylor also were discarded on empirical grounds. The defining characteristics of each stage simply did not hold up as anthropologists brought back information from cultures around the world. Many of the supposedly definitive traits assigned to the stages of savagery, barbarism, and civilization were found to occur in societies with markedly different levels of organizational complexity.

While mainstream anthropology in the early twentieth century may have disavowed the typological stage models of the previous generation, there was not a complete disavowal of the concept of cultural evolution. Scholars such as Robert Lowie (1920), a student of Boas and staunch defender of the particularist view of culture, clearly recognized and studied the developmental change of various aspects of cultural systems. Harris (1968) notes that Lowie's book

> *Primitive Society* is nothing if it is not a major contribution to the theory of cultural evolution. It is such a contribution because time and again in its pages Morgan's view of the sequence of the emergence of specific institutions on both a worldwide and more localized basis are examined, criticized, and reformulated with the positing of a new sequence. (Harris 1968:348; see also Lowie 1927)

Thus, even at the height of historical particularism in the first half of the twentieth century, there was recognition that human cultural systems have evolved over time and there were cross-cultural patterns to be seen in that evolution.

Although certainly dimmed in the first half of the twentieth century, the light of an explicitly evolutionary perspective was kept burning in the work of the British archaeologist V. Gordon Childe (1925, 1950). Childe was a persistent adherent to the stage model of Morgan; he also was the first to recognize that the study of cultural evolution was particularly compatible with archaeological analysis. He specifically attempted to reconstruct the archaeological sequences of Mesopotamia, Egypt and India in terms of the stages of savagery, barbarism and civilization. The significance of Childe's work, however, was not so much the use of evolutionary type categories as his explication of the process of cultural evolution over long periods of time. The preceding evolutionists of the nineteenth century had attempted to infer hypothetical evolutionary sequences by comparing different groups in the contemporary ethnographic record. Childe made a comprehensive effort to look at how prehistoric cultures actually changed over time—what were the steps involved in the move from one evolutionary stage to the next (Childe 1950).

BANDS, TRIBES, CHIEFDOMS, AND STATES

The explicit study and theoretical exposition of cultural evolution experienced an intellectual rebirth in the 1940s and 1950s with the work of ethnographers Leslie White and Julian Steward. Steward and White engaged in a lively intellectual debate about the nature and causes of cultural evolution. Steward (1951, 1955) proposed a model of "multilinear" evolution based on the concept of levels of sociocultural integration. According to Steward, "in the growth continuum of any culture, there is a succession of organizational types which are not only increasingly complex but which represent new emergent forms" (1955:51). The evolutionary levels of sociocultural integration included the "family," the "band"—patrilineal and composite— "folk society," and the "state." For Steward, these levels of integration were a typological means for analyzing cultures with varying degrees of complexity (1951:380), and to reflect the emergent organizational forms that recur cross-culturally in many different parts of the world.

Although Leslie White was a well-respected ethnographer of the Southwestern Pueblos (see White 1932, 1935, 1942, 1962), on a theoretical level his focus moved far beyond the evolution of particular cultural systems. White (1949, 1959) was primarily interested in the evolution of culture as a general phenomenon as opposed to the evolutionary development of specific cultures. While he certainly recognized different types of simple and complex social systems, he did not attempt to devise a typological system to group societies at different evolutionary levels. White spoke of "the culture of mankind in actuality is a one, a single system; all the so-called cultures are merely distinguishable portions of a single fabric" (1959:17). Looking at the whole of humanity, he presented a case that the basic function of culture is "harnessing of energy and putting it to work in the service of man" (1959:39). Accordingly, the evolution of culture in this model was defined in terms of the amount of energy captured and expended through technology and the economic system. Simple and undeveloped cultures are those in which energy is captured through human efforts alone. More complex cultural systems evolve as societies develop more effective means of harnessing energy through draft animals, irrigation, machines, and so forth. According to White, social and political complexity then evolved along with the progressive development of technological and economic systems for capturing energy.

While White and Steward were the clear leaders in the resurrection of cultural evolution studies in the mid-twentieth century, they saw their respective models as opposing rather than complementary. A subsequent generation of White and Steward students—led by Elman Service, Morton Fried, and Marshall Sahlins—sought to bring together the multilinear, *specific* evolution of Steward with the energy-based *general* evolution of White (Sahlins and Service 1960). They built on this specific/general foundation and developed alternative evolutionary models that looked at broad, *universal* patterns but could be applied to specific evolutionary processes in individual cultures.

The most comprehensive model of this generation was that of Elman Service, who rearticulated Steward and offered the evolutionary sequence of "band, tribe, chiefdom, state" (Service 1962; see also Sahlins and Service 1960:37). Service's model was derived directly from that of Steward in that he saw these sequential developmental stages as representing different levels of social integration. Band societies, according to Service, were relatively small, patrilineally organized groups of families engaged in a hunting and gathering subsistence economy. Tribes transcended bands by uniting groups of communities, held together by sodality relationships and practicing simple agriculture or pastoral nomadism. The chiefdoms in Service's model brought together a common population under the leadership of a central authority—a chief—and were based on the emergence of a redistributive economy. Finally, the state was a much larger polity with a centralized bureaucracy, engaged in large-scale agriculture and organized along territorial (rather than kinship) lines.

Morton Fried (1960, 1967), a close colleague of Service and also a student of Steward, developed an alternative evolutionary model that focused specifically on organizing principles of political organization. Fried's sequence of "egalitarian, ranked, stratified, and state society" is derived from cross-cultural comparisons of social status, access to resources, and the organization of power in different societies. Somewhat in contrast to Service, Fried was less concerned about the cultural or political composition of the stages and more concerned about the evolutionary transformations from one stage to the next. For example, how and why did cultures move from egalitarian, with equal social access to basic resources and positions of status to ranked society with equal access to resources but differential access to positions of status, and then on to stratified society with differential access to status and basic resources?

The evolutionary models of Fried and Service have provided a foundation for a tremendous amount of anthropological research in the course of the past 30 years. They have been widely applied in cultural anthropology both to examine aspects of social and political organization at different evolutionary levels and to classify societies into one of the evolutionary stages. The models also proved to be a bridge between cultural anthropology and archaeology as archaeologists found that the proposed evolutionary stages could be recognized in the archaeological record of past societies. The archaeological record, covering a long span of time, also proved to be a suitable laboratory for studying the actual process of cultural evolution as ancient cultures evolved from one stage to the next.

In the closing decades of the twentieth century, the stage models of both Service and Fried—like those of Morgan and Tylor before them—largely fell out of favor in both cultural anthropology and archaeology. As the Boasians did to Morgan and Tylor, the Service and Fried models have been criticized for overgeneralizing, obscuring cultural diversity, and inaccuracies at a number of levels. It has been found, for example, that many of the attributes bundled together in Service's different levels do not always (or even often in some cases) co-occur when examined in the ethnographic and archaeological records. Similarly, the hard and fast distinctions of power and status central to Fried's theoretical model turn out to be indistinct and transitional in the empirical record. There also is an underlying concern that behind all stage models, no matter how benign in appearance, there is an implicit notion of "progress" in the sense of advancement. The idea that the "state" or "civilization" is somehow the culmination of the evolution of culture is seen as reflecting a Western bias and does not give adequate regard to the accomplishments and independent character of nonstate societies.

The rejection of an evolutionary paradigm for studying culture broadly continues in contemporary cultural anthropology. The mainstream of the field has turned away from evolution to highly focused studies of individual cultures, culture traits and culture histories. With notable exceptions (see, for example, Ember and Ember 1994; Ferguson and Whitehead 1992; Johnson and Earle 1987; Fox 1991), there is little concern for or study of broad cross-cultural patterns of organization or the process of evolutionary transformation from one cultural configuration to another. In some ways, the current state of much of cultural anthropology harkens back to the era of Boasian particularism, when evolutionism was dismissed and anthropology concentrated on the details of cultures.

In archaeology, the study of cultural evolution has taken a different route. Archaeologists found that the great chronological depth of the archaeological record was particularly well suited to the study of long-term evolutionary change. Initially, archaeologists looked at the models of Service and Fried as a handy tool to organize prehistoric cultures into "meaningful" anthropological categories; they also used the stage definitions to flesh out the invisible details of the ancient past. Thus, for example, if an archaeologist found some evidence to indicate a particular culture was at Service's chiefdom level, it was inferred that the culture would have had all the attributes of Service's chiefdoms. However, as more researchers began to examine the organization and characteristics of past cultures, they increasingly found that many prehistoric societies did not have the full complement of traits ascribed to the different evolutionary stages (see Feinman and Neitzel 1984). Furthermore, the causal variables hypothesized to be driving the evolutionary process also did not seem to work as they were examined and tested in the archaeological record (see, for example, Bettinger 1991; Earle 1987a).

Rather than dismiss the evolutionary paradigm altogether, archaeologists have gone beyond the stage models and evolutionary theory of the cultural anthropologists to look for alternative strategies for studying the process of evolution (see Haas 1990; Earle 1991, 1997; Ehrenreich et al. 1995; Gregg 1991). Indeed, the intellectual responsibility for studying the evolution of complex cultural systems has shifted almost entirely to archaeology in the past two decades.

The study of cultural evolution from the perspective of archaeology represents an optimal wedding of subject matter and discipline. Although much of the early evolutionary theory arose out of cultural anthropology and the ethnological comparison of different contemporary cultures, the ethnographic record of individual cultures is necessarily limited to the present and at most one or two generations back into the past. To study the process of evolution ethnographically involves a circuitous inference of how one contemporary society might be representative of the evolutionary antecedents (or descendants) of another contemporary society. Such indirect inferences are difficult at best to test empirically and rely largely on the eloquence of logical argumentation. In contrast, the archaeological record by its very nature is diachronic and a more direct material manifestation of culture change over time. Archaeologists are able to study the remains of past cultural systems extending back over hundreds and thousands of years and to examine how those cultures evolved over time. The material record of past societies not only reveals *how* cultures evolved, but also provides the contextual information necessary to evaluate our ideas about *why* cultures evolved in response to environmental and social variables.

While archaeology is playing a predominant role in the study of cultural evolution today, other lines of research based in cultural ecology and sociobiology are looking in particular at evolutionary changes in behavior through group or cultural selection (Boyd and Richerson 1985; Boehm 1996; Soltis et al. 1995). Recent research in this area, borrowing heavily from biological evolution, turns to the ethnographic database to look at the mechanics of how cultural information and innovations are transmitted across cultural lines and accordingly how cultures may evolve through such mechanics. These studies can potentially supplement and expand archaeological studies by providing specific detail on actual evolutionary processes at the level of families, communities, and culture groups. However, there has been little application of such approaches to the archaeological record of cultural evolution.

SELECTION AND TRANSFORMATION

The shift of cultural evolutionary studies from ethnology to archaeology has been accompanied by new bodies of theory and alternative perspectives on the evolution of cultural systems. Although there are many different viewpoints on the archaeology of evolution, most work can be broken into two general schools of thought: the selectionist and the transformational. The *selectionist* school, championed by Robert Dunnell and his students (Dunnell 1980, 1989; Leonard and Jones 1987; Teltser 1995), has intellectual antecedents in both the particularism of Boas and the specific evolution of Steward. The current school takes a Darwinian or neo-Darwinian approach to the evolution of culture. The basic premise of this approach is that there is a range of variability to be found in any culture and that evolutionary change occurs through selection for some part of that variability over some other part. David Braun has neatly summarized the basic selectionist model:

> We should approach social evolution as a matter of "descent with modification," to use Darwin's phrase. But by "descent" here I refer to sociocultural rather than biological descent. That is, we should analyze social evolution as a matter of continuity and change in the statistical popularity of different social practices over time among individual human communities. . . . We can describe continuity and change in the popularity of different social practices in terms of two phenomena (e.g., Dunnell 1980, 1989). First, there always exists variation in the statistical popularity of different social practices at any given time among different interacting sets of people. . . . Second, there always exists a pattern of differential transmission of that variation over time. (Braun 1990:63–64)

The selectionist approach in archaeology, because of its emphasis on the evolution of particular attributes within specific cultures, has been used primarily as a tool for explaining variation in the prehistoric past:

> For more than fifteen years a variety of archaeologists have advocated the application of Darwinian evolutionary theory to explain variation in the archaeological record.... In short, expanding evolutionary theory to explain variation in the archaeological record requires building new archaeological theory and method. (Teltser 1995:1)

A contrasting *transformational* view of culture is intellectually descended from the earlier stage models of cultural anthropology and from the general evolution of White. There are two fundamental principles to this view: (1) in the course of evolution, cultural systems are transformed from one organizational structure to another; and (2) there are broad cross-cultural and cross-chronological patterns in the evolution of cultural systems around the world.

The transformative aspect of cultural evolution is based on the premise that not all change is gradual and quantitative. Some changes result in qualitative transformations of entire cultural systems (see Haas 1982; Kristiansen 1991; Price and Feinman 1995; Wason 1994). The emergent, transformed system is not just a minor variation on the antecedent system, but something that is fundamentally new and different from previous forms. For example, individual households in an area may aggregate together into villages for a variety of reasons. The resulting villages, however, are more than just physical aggregates of households. They represent a fundamental transformation of the areal settlement system requiring new organizational principles, new forms of social relations, and significant shifts in resource procurement strategies (see Kohler and Van West 1996). This kind of qualitative transformation can be observed in almost every dimension of human cultural systems, including social relations, politics, art, religion, economy, war, technology, and so forth. In the context of this volume, the emergence of political centralization—while it may have occurred in a series of small steps over time—represented a pivotal transformation of social relations. A politically centralized society is not just an egalitarian society grown bigger and more complex; it is a profoundly different kind of society from its egalitarian evolutionary antecedents.

Coupled with the qualitative nature of certain kinds of evolutionary changes are broad patterns to the transformational processes that cross cut time and geography. In the case of social organization, political centralization did not arise in just one location and then spread inexorably around the world. Rather, centralized polities have arisen independently at different times in many parts of the world over the course of the past 10–15,000 years. Although each instance of centralization is historically unique in terms of specific circumstances, there are nevertheless strong patterns and similarities that pertain to all or most of these historical sequences.

Beyond patterning in social organization, broader crosscutting patterns can be observed in the full range of evolving human societies subject to transformational change (see DeMarrais et al. 1996.; Earle 1998; Ames 1991; Paynter 1989; Blanton, et al. 1996). Settled village agriculture with similar kinds of social and political institutions, for example, independently develops in many different parts of the world over the course of several thousands years in human history. Subsequently in a number of widely disparate parts of the world (Asia, Africa, South America, Europe, and Polynesia) we see the independent development of centralized societies with social ranking led by "chiefs." Then in different centers of world civilization (Mesopotamia, Egypt, China, India, the Andes, Mesoamerica, and West Africa) there is the independent development of remarkably similar bureaucratic polities with marked social stratification, cities, elaborate art, and centralized religion. Empires follow in some world areas as do feudal societies and the nation-states of the recent historical record. The transformational model recognizes these broad patterns found in the evolutionary trajectories of many cultures and looks for explanations that transcend the specifics of individual cases.

In contrast, then, to the selectionist focus on variation, the transformationist focus is on understanding the pattern and processes inherent in these cross-cultural transformations of cultural systems. The focus on pattern and process in transformational approaches does not automatically preclude the study of cultural variation. Thus, it is possible to examine the evolutionary transformation from decentralized autonomous villages to centralized polities in one particular area as a specific subset of the cross-cultural pattern of political centralization found in many areas at different points in time.

Although selectionist and transformational models are sometimes seen as standing in stark opposition to one another (Leonard and Jones 1987; Braun 1990), acceptance of one need not lead to the exclusion of the other. The two perspectives actually complement each other in interesting ways as alternative tools for studying the evolution of culture as a complex adaptive system. The selectionist model has largely abandoned the quest to address general patterns of evolution that recur cross-culturally around the world and across long spans of time. It emphasizes instead the importance of being able to understand the unique evolutionary trajectory of individual cultures and focuses on the specific mechanics of those evolutionary events. Through this approach, valuable insights can be gained into the social mechanics of change at the local level. At the same time, there are remarkably similar social and cultural phenomena that crop up again and again around

CULTURAL EVOLUTION AND POLITICAL CENTRALIZATION

the world and these similarities, these patterns of evolutionary change, constitute grist for the mill of transformational approaches.

Certainly there are individual circumstances that make each evolutionary change a unique historical event and there are singular evolutionary forces operating on all cultures at any given point in human history. At the same time, people of different cultures respond to their unique circumstances in similar ways with a relatively tightly defined range of variability. Explanatory models that can account for the similarities of evolutionary change across cultures and across time can greatly extend our insights into patterns of human behavior, both past and present.

TINKERING AND TRAJECTORIES

The question then becomes how to combine the two evolutionary approaches in a more comprehensive effort to understand both the mechanics of cultural evolution at the local, societal level and the kinds of cross-cultural trajectories that can be seen to recur around the world over the past 100,000 years of modern human history. One possible avenue for looking at both the mechanics of specific evolution and the patterns of general evolution is through a more explicit consideration of human agency—the actions of individual human actors in a social system—and what I would like to refer to as the "tinkering" view of evolution.

Actually, the idea of tinkering has historical roots in evolutionary biology and in general anthropology. Francois Jacob (1977) wrote about the overall process of evolution being like the actions of a tinkerer: "Evolution behaves like a tinkerer who, during eons upon eons, would slowly modify his work, unceasingly retouching it, cutting here, lengthening there, seizing the opportunities to adapt it progressively to it's new use" (1977:1164).

However, there are intellectual antecedents to Jacob's tinkerer in the work of anthropologist Claude Lévi-Strauss (1966) writes about "bricolage" in a similar fashion to discuss how new myths develop out of an existing repertoire. The bricoleur or tinkerer works with what is readily available at the moment to devise new tools and solutions to problems:

> His [the bricoleur's] universe of instruments is closed and the rules of his game are always to make do with "whatever is at hand," that is to say with a set of tools and materials which is always finite and is also heterogeneous because what it contains bears no relation to the current project, or indeed to any particular project, but is the contingent result of all the occasions there have been to renew or enrich the stock or to maintain it with the remains of previous constructions or destructions. (Lévi-Straus 1966:17)

The application of the tinkering concept here differs from these intellectual antecedents in assigning the role of tinkerer to all the individual human agents in a cultural system. The basic premise of cultural tinkering view is that in every society across time people are and always have tinkered with their world. Farmers tinker with they way they plant, weed, and harvest; weavers tinker with yarns and warps; storytellers tinker with their stories; politicians tinker with politics. In the selectionist model, the process of evolution occurs through selection for or against the variation resulting from cultural tinkering (Braun 1995). To the extent that such tinkering then survives a process of cultural selection (see Cavalli-Sforza and Feldman 1981; Boehm 1996, 1997; Teltser 1995; Boyd and Richerson 1985) it may be integrated into the standard cultural repertoire.

While it can be expected that this kind of tinkering by agents goes on under any and all cultural conditions, it also can be expected that the level and direction of tinkering will be affected by the social and environmental conditions that may be influencing any given culture at any given time. This is little more than a restatement of the adage "necessity is the mother of invention." Thus when harvests are good and there is ample food for all, it can be expected that the level of tinkering of farmers will be relatively low. In contrast, when harvests are bad and people are going hungry, it can be expected that there will be a relatively high level of tinkering. Some of this tinkering may yield better results and some will not. This pattern is amply illustrated in the archaeological record where long periods of relative abundance are marked by cultural stability and slow, accretionary cultural change. In contrast, periods of scarcity—manifested in nutritional stress—witness much more rapid change and transformation of old patterns (in the American Southwest, for example, see Adler 1996; Gumerman 1988; Haas 1989).

While most innovations introduced through tinkering will have relatively small impact on the cultural systems, some innovations will have much broader, transformational consequences. Successful tinkering in one part of a cultural system may also lead to rippling changes in other parts of that system as accommodations are made for new practices, ideas, or relationships. Thus, for example, the decision of a farmer to plant three seeds in a hole instead of two may ultimately yield more grain but have few far-reaching effects in the organization of the culture. In contrast, the initial small-scale decision to channel water away from a river to irrigate adjacent fields could have a rapid and dramatic impact not just on crop yield but on the organization of the entire culture (Downing and Gibson 1975).

On the surface, the principle of tinkering would appear to be anathema to the effort to extract and understand pattern and process in the evolution of cultural systems. If change comes about as a result of selection among the tinkering activities of all the agents in a population, we might expect to find an infinite unordered range of variation in the record of past

CULTURAL EVOLUTION AND POLITICAL CENTRALIZATION

human societies. In fact, we do find this infinite variation in the unique histories of every human culture, but underlying this variation patterns emerge as groups of people across time and space converge on similar solutions to similar problems. To draw an analogy from another field, the relationship between tinkering and patterning is nicely illustrated in the movie industry. Every movie is a unique product of the creative talents of the screenwriter, the director, the camera person, the actors, and so on. Some movies are deemed good, others are bad; some succeed at the box office, others fail. Out of this process of creative tinkering, clear patterns have emerged in the course of American cinematography: the good guy wins, crime doesn't pay, boy gets girl/boy loses girl/boy gets girl. Of course in the movies, as in culture, there are exceptions to these patterns as the agents involved tinker with alternative plots, characters, and conclusions.

Applying the principle of agent-based tinkering to the evolution of culture allows for the combination of basic principles of selectionism—generation of variation and selection of advantageous options—with the quest to understand patterns of transformation within cultural systems. It necessarily takes the focus away from broad stages, which are defined in terms of entire cultural systems, and places the focus on separate parts of systems where agency can be more effectively incorporated into explanatory models. Thus, for example, a consideration of the role of human agency and tinkering may not provide useful insights into why "tribes" evolve out of "bands," but it can be very useful in shedding light onto parts or aspects of that transformation. Thus, for example, the decision of people to agglomerate into villages may be seen in any given historical sequence as an example of local-level tinkering. It is one option people may pursue in an effort to resolve particular problems such as threats from enemies, need for communal storage, enhanced communication, or localized water resources. Other people in that same area or region may not choose to move into villages and retain their isolated farmsteads. Depending on the relative success of the aggregation strategy/option under the particular historical circumstances, of the time the village pattern may or may not spread throughout the region. Based on global records, the village option was clearly successful and selected by small groups of people at many different times and places. Furthermore, the pattern of village formation also is closely correlated with other kinds of patterns such as warfare, agricultural intensification, and increased social integration.

Although the monolithic state models (band, tribe, chiefdom, state; egalitarian, ranked, stratified, etc.) have played an important heuristic role in highlighting high-order patterns in human history, they do not necessarily provide the most effective vehicle for understanding the wide range of patterns manifested in the long term evolution of cultural systems. Extensive anthropological research over the past two decades has shown that many of the attributes—demographic, social, political—thought to be "coupled" in discrete stages in fact are "decoupled" in many societies (see Upham 1990; Earle 1991). The interconnections between disparate cultural traits are proving to be much more complex than was envisioned when the stages were initially defined.

A more productive strategy for approaching cultural evolution can be found by shifting the focus away from monolithic stage concepts to broad trajectories of change that may be seen as subcomponents of the process of cultural evolution. Although there is potentially an infinite number of cross-cultural patterns that have evolved around the globe, a workable number of broad patterns have been both widely recognized and subjected to extensive anthropological study. The term "trajectory" is used here explicitly to connote the path of development that can be observed cross-culturally in the evolution of different aspects of complex cultural systems.

- *Development of technology.* This is basically the trajectory that Leslie White (1949, 1959) identified as the increased harnessing of energy. In agriculture, mining, harvesting, and other areas, humans have developed technological strategies for more efficient and effective extraction of resources from the natural environment. In some cases, such as the transition to an agriculturally based economy, technological evolution has had a profound impact on all quarters of the cultural system. (Layton et al. 1991; Wills 1988; Smith 1992).
- *Increasing settlement aggregation and the rise of cities.* As recently as 15,000 years ago, virtually all of humanity lived in relatively small, mobile bands of hunters and gatherers. Over the millennia, people have come to settle in ever-larger villages, towns, and cities up to the megalopolis of the modern era (Ames 1991; Possehl 1990; Kolb and Snead 1997).
- *Alliance formation and increasing social integration.* As culture has evolved, there has been a trajectory toward forming more formal social alliances between different political, residential, and kinship units. The nature of social integration holding the diverse parts together also has become more complex over time in many cultures (Gregg 1991; Upham 1990).
- *The development of social hierarchies.* There has been a general trend from relatively egalitarian, nonhierarchical societies through various stages of social differentiation through the development of caste and class societies (Paynter 1989; Wason 1994; Price and Feinman 1995; see also Ehrenreich et al. 1995).
- *Centralization and the evolution of power.* The trend here has been

toward both increasing centralization of power held by some segments of a society and the differentiation of power holders dependent on economic, ideological, physical, and political bases of power (Joyce and Winter 1996; Schortman and Urban 1992; Patterson and Gailey 1987).
- *Craft specialization and increased division of labor.* In early hunting and gathering societies, there is little specialization and a relatively simple division of labor based on age and sex. Subsequent evolution has involved both increased specialization in production of goods and services as well as division of labor along many lines beyond age and sex (Brumfiel and Earle 1987; Marshall 1990).
- *Ethnic differentiation.* There has been a consistent global trend toward increasing separation of ethnically diverse groups over the course of the past 100,000 years (Sampson 1988; Green and Perlman 1985; Neiman 1995; Shennan 1989; Jones 1997).
- *The emergence and evolution of war.* War, as organized violence between political units, appears at different times in different cultures around the world. Once it has developed, the nature of war also changes in patterned ways cross-culturally (Keeley 1996; Otterbien 1985; Haas 1998).
- *The formation of multiethnic polities.* Earlier polities, from tribes to states, all tend to be ethnically homogeneous in that the people speak the same language, have similar customs, material culture, and so forth. Later polities develop with several or many different ethnic groups living together under a single centralized government (Hassig 1988; Schreiber 1992; Sinopoli 1994).

There is no implication in any of these trajectories that evolution somehow involves "progress" in any subjective or empirical sense. In all of them, however, there are clear patterns that recur in the development sequences from early to late in many societies. These are the patterns that demand the attention and explanation of anthropology.

All these evolutionary trajectories are internally complex and none are inevitable universals. There can be divergences and even reversals in any of the general trends and different cultures will have their own particular manifestation of each element. There are certainly some linkages between all of these, though again the connections are not inevitable and globally universal. (Thus, for example, social hierarchies may develop in the context of agricultural intensification in many societies, but in the absence of agriculture in others.) Nevertheless, these different trajectories each represent a separate line that can and has been examined cross-culturally and across time to extract and explain common paths of evolutionary change. It also is possible to use modeling to look at the intersections of the different trajectories and examine how they may be connected or disconnected in *both* individual culture histories *and* broad patterns of evolution.

The chapters in this volume represent an effort to focus on political centralization and the evolution of power. They represent an attempt to tease out both similarities and differences along the broad, cross-cultural trajectory toward increased centralization and concentration of power in the hands of ruling elites. Collectively, they illustrate the cross-cultural patterns manifest in the process of political centralization; individually, they demonstrate that centralization is not a monolithic trajectory but historically and geographically contingent on local and regional variables. They also show that centralization must be assessed at multiple levels in complex polities. While a society may be strictly hierarchical and centralized at one level, it may be heterarchical and decentralized at another. In general, the larger and more complex a polity, the more likely it is to incorporate a combination of centralized and decentralized elements. At the same time, it must be recognized that as power differentials increase and political systems generally become more complex, decision-making authority comes to be increasingly concentrated in the hands of the political elite.

In the following chapter, Crumley, challenges all of our existing models of cultural evolution. Crumley does not summarily dismiss the work of her predecessors but argues instead that we have now learned enough to move in new directions and think in new ways. The chapter represents a truly remarkable effort to integrate alternative theories of causality with environmental and cultural data drawn from across the globe. Her thoughts, while aimed at political organization, cut across the entire spectrum of long-term cultural change and ultimately provide a new way of looking at processes of tinkering and transformation.

Chapter **2**

Communication, Holism, and the Evolution of Sociopolitical Complexity

CAROLE L. CRUMLEY

INTRODUCTION

In the eighteenth century, Europeans embraced church doctrine that ranked all creatures: the Great Chain of Being offered a hierarchy of divine order that both upheld class distinctions at home and served as a pious rationale for colonialism. By the end of the nineteenth century the new sciences of society had replaced religious doctrine, but scientific thought, drawing on the idea of progress, upheld similar conclusions: society had evolved "naturally" toward increasing inequality and complexity.

Today, at the beginning of the twenty-first century, every term and link in this formulation is being reexamined. The ideas of progress and nature have proven suspect, evolution appears to lurch rather than power glide, the biological analogy needs sober rethinking, and the search for a flexible definition of complexity is everywhere underway.

The biological and social sciences share a long history of reciprocal borrowing and legitimation regarding theory (Ellen 1982). One shared theme

---

CAROLE L. CRUMLEY • Department of Anthropology, University of North Carolina—Chapel Hill, Chapel Hill, North Carolina 27599.

From Leaders to Rulers, edited by Jonathan Haas. Kluwer Academic/Plenum Publishers, New York, 2001.

tests the limits of the applicability of Darwinian analogy. A second theme, termed self-organization, explores the informational context in which evolution takes place. A third explores the relationship between hierarchy and complexity. These interdisciplinary theoretical trends are pertinent to many fields of endeavor, including anthropology, sociology, biology, physics, computer design, artificial intelligence, ecology, business, education, and religion. This chapter links three central themes (evolution, self-organization, and complexity). Together they offer a fresh perspective on state formation and dissolution.

EVOLUTION, SELF-ORGANIZATION, AND CHAOS

Charles Darwin's vision of how change occurs in living organisms is being seriously challenged for the first time since its general acceptance. To be sure, the concept of evolution, despite some tinkering with issues of timing by means of the concept of punctuated equilibrium, still has abundant evidential affirmation. The challenge is not that evolution does not occur, but that it is not sufficient to explain the complexity of the universe.

Since its founding in 1984, researchers at the Santa Fe Institute have concentrated on understanding complexity. They and investigators elsewhere have a new candidate idea: self-organization. Briefly, they argue that the introduction of transmission errors through mutation and the operation of selection do not alone explain the complexity that may be seen in myriad living systems, from fireflies to fiddle music. They assert that evolution forces us to see a universe in which randomness alone explains the infinitesimal chance that life could be created out of a chemical soup.

They argue that a second, more fundamental source of order exists, called self-organization. This means that there is a synergy that comes from communication, and that two (or more) communicating entities have different properties than each alone or their noncommunicating kin. Groups of communicating organisms, if sufficiently diverse, can become self-sustaining, that is, alive; this is termed *autopoesis*. Thereby, they may be transformed into a more complex system, the stepwise evolution of which is termed *anagenesis* (Jantsch 1982:345; Kauffman 1995:64). Autopoesis also is referred to as receiver-based communication, where all agents report to other agents what is happening to them (Kauffman 1995:267). The human body is an example of autopoesis; the evolution of the human mind is an example of anagenesis (Mithen 1996).

Educators Fleener and Pourdavood (1997:10) note that the development of communication is important for both the emergence of cognition in human history and the coupling of humans within a social domain. They relate autopoesis—the self-renewing, autonomous, reproductive aspect of self-organization—to two levels of communication: language and social structure. That relation persists in the form of social memory (McIntosh et al., 2001), which is collectively stored and passed on from generation to generation (Crumley 2001; Gunn 1994). This is, of course, an essential definition of culture.

The control of information in complex polities has been the subject of considerable research (e.g., Johnson 1982), but the focus has been on the flow of information to decision makers, rather than in the degree of connectivity among all elements of the population. While control over information is always a source of individuated power, the study of self-organizing systems demonstrates the utility of a more inclusive and social definition of communication.

The governing assumption in self-organization research is holism, the idea that an organism is more than just the sum of its parts. The self-organization researchers are critical of the past three centuries of positivist scientific endeavor, where the basic assumption has been that if the entity (living or not) can be broken down into its constituent parts, its behavior can be understood.

The current generation of scholars pursuing the "science of complexity" do not advocate the abandonment of Darwinian evolution as a central paradigm, but rather the addition of self-organization. Together, they argue, selection and self-organization form the structure of the universe; neither alone suffices. Together, Darwinian evolution and self-organization bring order from chaos: self-organization creates new forms and evolution judges their goodness of fit. Each new stage of organization has the potential for further evolution (De Greene 1996:276), here taken to mean the transformative nature of communication.

Related research has explored *chaos*. Chaos is a condition and implies the existence of unpredictable aspects in dynamic systems. This is not necessarily undesirable, as it also is a source of creativity; as the American essayist and historian Henry Adams put it, "Chaos often breeds life, when order breeds habit" (Çambel 1993:15). Far from exhibiting the absence of order, chaotic systems have structure and operate within demonstrable parameters.

Antecedent conditions always affect system parameters (thresholds), and thus system stability. System parameters can be known but they cannot always be predicted. If we envision order to be predictability, then chaos would be surprise (Kauffman 1995:15). Systems in equilibrium (such as a ball in the bottom of a bowl) are low energy; it is the high-energy nonequilibrium but nonetheless ordered systems (such as the human organism or the governance of a complex polity) that we seek to understand.

Such systems invariably operate at the edge of chaos, that is their behavior is quite unpredictable, but not entirely so. Chaos researchers think the key to predictability lies in understanding *attractors*, which are sources of order in large dynamic systems (Kauffman 1995:79). Attractors are the form, or state, that a cycle of change takes and that the system periodically "visits."

For example, consider a pendulum suspended from a height and moving through space and in time. We can know the rules that govern its movement and have high expectations of its predictability. This is called a *fixed-point attractor*. A slightly less predictable attractor is called a *limit cycle*, of which predator-prey relationships are an example. A *torus attractor* is useful in thinking about systems with many degrees of freedom; an example here would be two predator–prey relationships that are connected. Finally, a *strange attractor* would characterize complex aperiodic systems, like a whirlpool or a hurricane or perhaps states. Strange attractors are highly irregular but crucial to dissipating dynamic systems and are associated with chaos. In dissipative systems, the flux of matter and energy through the system is a driving force generating order (Kauffman 1995:21). The different attractors are related, in that each demonstrates pattern and regularity, even strange attractors (Çambel 1993:69). The phase–space in which the attractors operate is called their *basin of attraction*.

A means of understanding systemic parameters called rule-based learning has been successfully employed in the study of global climate, a complex, dissipative system which has both long- and short-term impacts on human society (Gunn and Grzymala-Busse 1994). The LERS (learning from examples based on rough sets) program, termed the "leave-one-out test," systematically leaves the global average temperature for each of a series of years out of the calculation, regenerates the rules with each year missing, and determines whether it can identify the climate of the removed year.

The study identifies three attractors (forms or states that a cycle of change takes) that the global climate system revisits with variable periodicity: hot, cold, and moderate. The study also explores three principles: trends, canceling roles, and reinforcing roles, which determine the nature of each climate state (basin of attraction). Rule-based learning offers an effective means by which the history and future of two complex systems—global climate and human societies—can be understood.

Another way of thinking about the relationship between evolution and self-organization is that they form a *dialectic*, which suspends the utility of old forms because of new, unexpected, conditions (chaos). Then, drawing on a fund of diversity (mutation), preserves some old elements of those forms and discards others [selection], then transcends the previous form (communication results in emergent properties), thus creating a new, more complex form (a new interface between order and chaos).

Such an understanding of human history is profoundly nonlinear, while remaining holistic and cumulative. It appropriately reminds us of the need for balance: the more complex (i.e., densely connected) the system, the more likely it is to be unstable, yet systems at the edge of chaos have order and are flexible and creative.

Much of the argument just outlined is familiar to anthropologists, who take the view (also termed "holism") that culture is greater than the sum of its constituent parts. The central importance and the transformative nature of communication, which in its diversity is undeniably one of the hallmarks of human society, also is welcome. The idea is that new forms and relations ("emergent properties") come into being when information is shared, allowing entire systems to be transformed (to change their state). As Goodwin (1994:xiv) has put it, "It is relational order *between* components that matters more than material composition in living processes, so that emergent qualities predominate over quantities. This consequence extends to social structure, where relationships, creativity and values are of primary significance."

Finally, the idea that initial conditions shape system history has reintroduced the importance of time to several disciplines. Archaeologists in particular have long realized that all human societies are constrained by their location in time and space; the work on initial conditions supports the importance of history, as event (*evenément*), as context (*conjoncture*), and as process (*longue durée*) (Burke 1990).

The research in self-organization represents an historic break with the discipline-based, ahistorical, and reductionist agenda that has characterized social and physical science research since the mid-twentieth century. To the extent that holism, communication, and history are central to the anthropological enterprise, the central tenets of self-organization research are compatible.

Working in perennially treacherous interpretive terrain and with partial evidence, archaeologists have long searched for theory sufficiently robust to bridge the many research domains they employ. For most of this century, archaeologists labored to fit recalcitrant data into the dainty shoe of Darwinian evolution (e.g., Dunnell 1989; Soltis et al. 1995). Nonetheless, much remains unexplained, and discomfort with the evolutionary paradigm grows (Ehrenreich et al. 1995; Chapter 1, this volume).

Why is the brilliance of one society soon eclipsed while another enjoys long-term success? How does a society's past limit its current choices? How has environmental change affected the longevity of state societies? Are some forms of organizational structure more suited to particular conditions than others? It is clear to many archaeologists that the evolutionary paradigm alone cannot address these issues. Self-organization and related concepts, however, can transform the current theoretical framework without losing the progress that has been made.

ORGANIZATIONAL COMPLEXITY: HIERARCHY AND HETERARCHY

Human organization, by measures of adaptability and interactivity, is arguably the most complex category of self-organizing system known. From the earliest human societies to the present day, individual creativity and flexibility with regard to ecological niches have met with success. The toleration of difference in individuals and of variety in circumstances increases societal choice and offers a reserve of alternative solutions to problems. Organizational flexibility—in economic, social, and political realms—enables societies to adjust to changed circumstances.

Anthropologists attribute considerable flexibility to bands and tribes, but much less to stratified society (chiefdoms and states). Yet although hierarchical organization characterizes many aspects of state power, hierarchy alone does not capture the full range of state organizational relations. Heterarchies of power—coalitions, federations, leagues, associations—are just as important to the functioning of many states, not just so-called egalitarian groups (bands and tribes).

Hierarchy (the classic organizational pyramid commonly found in business and government) is a structure composed of elements that on the basis of certain factors are subordinate to others and may be ranked (Crumley 1979:44, 1987a:158). In a *control hierarchy* each higher level exerts control over the next lower level; the US court system and the army are control hierarchies. By contrast, disturbances at any level in a *scalar hierarchy* can affect any other scales (Crumley 1995a:2). This is because in control hierarchies, individuals and groups with authority and those with responsibility are isomorphic; communication becomes a commodity to be hoarded. In scalar hierarchies, for better or worse, elements at all scales are in communication with elements at all other scales. Following Mingers (1995), Fleener and Pourdavood (1995) evoke a scalar hierarchy of systems, ranging from the simplest (static, mechanical) to the most complex (language, self-consciousness). This is the same argument advanced by Jantsch (1982:349), for all scales must and do make up a holistic world.

Heterarchy is the relation of elements to one another when they are unranked or when they possess the potential for being ranked in a number of different ways, depending on conditions (Crumley 1987a:158). Power, understood from a heterarchical perspective, is counterpoised and linked to values, which are fluid and respond to changing situations. This definition of heterarchy and its application to social systems is congruent with Warren McCulloch's (1945) research into how the brain works.

It was McCulloch, a strong influence on the self-organizing systems theorist Kauffman (1995:xx), who first employed heterarchy in a contemporary context (McCulloch 1945). He examines independent cognitive structures in the brain, the collective organization of which he terms "heterarchy." He demonstrates that the human brain is not organized hierarchically but adjusts to the reranking of values as circumstances change.

For example, someone may highly value human life in general but be against abortion rights and for the death penalty (or vice versa). The context of the inquiry and changing (and frequently conflicting) values (Cancian 1965; Bailey 1971; Crumley 1987b) mitigates this logical inconsistency and is related to what Bateson (1972) terms a "double bind." Priorities are reranked relative to conditions and can result in major structural adjustment (Crumley and Marquardt 1987:615–617). McCulloch's (1945) "nervous nets," the source of the brain's flexibility, is a fractal (same structure at a different scale) of the adaptability of fluidly organized, highly communicative groups.

McCulloch's (1945) insight about the autonomous nature of information stored in the brain and how parts of the brain communicate revolutionized the neural study of the brain. It also solved major organizational problems in the fields of artificial intelligence and computer design (Minsky and Papert 1972). What McCulloch realized was that information stored in bundles as values in one part of the brain may or may not be correlated with information stored elsewhere, depending on the context; in computer terminology, subroutine A can subsume ("call") subroutine B and vice versa, depending on the requirements of the program. Rather than the "tree" hierarchy of the first computers, those today use an addressing (information locating) system that is heterarchical.

In summary, heterarchies are self-organizing systems in which the elements stand counterpoised to one another. In social systems, the power of various elements may fluctuate relative to conditions, one of the most important of which is the degree of systemic communication.

The addition of the term "heterarchy" as a descriptor of power relations in complex societies (Crumley 1979, 1987a, 1995a) is a reminder that there exist forms of order that are not hierarchical and that interactive elements in complex systems need not be permanently ranked relative to one another. Heterarchy describes the relation of elements to one another when they are unranked or when they possess the potential for being ranked in a number of different ways. In attempting to maintain a permanent ranking by consolidating power and melding rigid sociopolitical and religious hierarchies (hyperhierarchy or hypercoherence), the necessary flexibility in dealing with surprise is lost.

AUTHORITY STRUCTURES: HIERARCHIES AND HETERARCHIES

White (1995:118) provides a useful scheme for understanding continua in the various organizational dimensions of complex societies. For both hyperhierarchical states and those more heterarchically organized, White characterizes individual rules for behavior, gender relations, economy, social status, conflict resolution, social ideology, the political relation between leaders and followers, and temporal dynamics. To this I wish to add an examination of the contrasting conditions of decision making and clarify a single link: the relation between administrative structure and environmental stability and change.

Hierarchical Polities

Administrators in strong hierarchies (hypercoherent authoritarian states termed "hyperhierarchies") have the following advantages. Due to a clear decision-making chain, they respond well to fast-developing crises (e.g., military attack, insurrection). Because the rules and responsibilities are known to all, political interactions among decision makers are few and formalized, and political maintenance of the system is low. Administrative hierarchies are equipped with powerful security forces that can successfully defend the state perimeter and suppress internal dissent.

Hierarchical polities are at a disadvantage because data-gathering techniques, tied to the pyramidal decision-making framework, slow the arrival of some kinds of information (especially subversive activity) at the apex of the pyramid and necessitate the formalization and elaboration of internal security forces. Decisions are rapid and expedient but they are not necessarily popular; popular dissatisfaction is high and there must be considerable investment in coercion and/or chicanery. In any event, security costs are high.

Heterarchical Polities

Administrators in heterarchically organized polities are treated to good-quality information from many sources within and outside of the decision-making lattice. For the most part, decisions are fair and reflect popular consensus. Decision makers hear of a variety of solutions to problems. Because heterarchies are more likely to value the contributions of disparate segments of the community (women, ethnic groups, etc.), the society as a whole is better integrated and the workforce is proud and energized.

Heterarchical polities are at a disadvantage because consensus is slow to achieve, increasing the time it takes to make a decision (but see below). Decision makers must engage in interpersonal dialogue with constituents, which requires considerable time and energy investment and constant maintenance. The cacophonous voices and choices a decision maker hears render difficult the search for workable solutions.

EVOLUTION OF SOCIOPOLITICAL COMPLEXITY

Trade-offs

The greater the group involvement, the greater the range of response choices and the more inclusive the consensus, but the response time is slower and long-range planning is more difficult. Spontaneity, polyvalent individuality linked to achieved status, inclusive or counterpoised definition of state power, and flexibility are valued in heterarchies; rule-based authority, rigid class lines linked to ascribed as well as achieved status and rank, a control definition of state power, and the status quo are valued in hierarchies. Of course, state democracies exhibit characteristics of both, which explains in part why they are more stable than authoritarian states.

HETERARCHIES IN BUSINESS: MATRIX ORGANIZATION

How can we know more about how heterarchies function? Fortunately, the business community has been experimenting with heterarchical, or matrix, organizations since the 1960s (Anderla et al. 1997; Stark n.d.; Wheatley 1994). Today, matrix organization is widely accepted, especially in the international business community (Bartlett and Baber 1987; Benedetto 1985; Business International Corporation 1981; Cleland 1984; Davis and Lawrence 1977; Hill and White 1979; Janger 1963, 1979; Kerzner and Cleland 1985; Knight 1977; Kramer 1994; Morrill 1995; Wheatley 1994). Matrix organizations are characterized by a multiple-actor command structure, and shared resources and power; they foster appropriate support mechanisms, organizational culture, and behaviors.

As Japanese companies gained market share in the 1970s, the traditional Japanese *kieretsu*, which are networks of industrial, transportation, and financial "societies of business," were closely analyzed by the Western business community. The first American matrix organizations, loosely organized around some of the kieretsu features, emerged in response to the then-novel organizational requirements of the aerospace industry; as global markets became commonplace in the 1990s, these power and resource matrices also became multivalent ones of scale and value.

They work as follows: Flexible *ad hoc* teams undertake multiple and changing projects, and cultivate diverse clients and markets. Management authority structures (whose knowledge is valued?) and responsibility structures (who makes sure the job is done?) are disconnected from one another; thus, the shop foreman's experience can influence the restructuring of the assembly line). Manager/facilitators are connected through meetings with those who implement decisions through their work. Because of their emphasis on communication, matrices cannot be simply installed in the workplace but need to grow over a period of time (Davis and Lawrence 1977).

Now that matrix management principles have been implemented in government and business settings for more than two decades, the "strengths-and-weaknesses" analysis of their application is particularly valuable to students of the hierarchy–heterarchy relation. For example, recent research in evolutionary economics and (corporate) organizational analysis suggests that features that retard the quick pursuit of immediate success (e.g., elements of institutional history, a lengthy decision-making process) keep alternative courses of action viable. Friction (such as that caused by differing knowledge, history, or values) can, in times of change, favor organizations that excel in the search for new approaches (Stark n.d.:5). Heterarchical organizations, characterized by diversity and broadly communicating elements, would be at an advantage in uncertain times.

LIMINALITY

Self-organizing systems are able to perform the most sophisticated computations when operating at the boundary between order and randomness (Langton 1992). By measures of adaptability and interactivity, human societies are arguably the most complex category of self-organizing system known. Individual creativity (cognitive liminality) and ecological flexibility (adaptability to varied ecosystems and to ecosystemic change, strategic positioning of communities at the juncture of biotic zones) are hallmarks of human societies at all times and places.

The concept of self-organizing systems leads me to suggest that human adaptive success is related to the juxtaposition of ecological and cognitive liminality (Ellen 1982; Turner 1995) with flexible (heterarchical) power relations. Humans are an ecotonal species, as evidenced by our beginnings on the forest–savannah margin and by our millennial fever to build and connect despite clear risks. While creativity and innovation are often associated with individuals and groups at the margins of social space (Barnett 1953), it is also there that they pose the greatest danger to the state. Hence the always uncomfortable and often hostile relation between authoritarian states and the arts.

A key component of belief systems worldwide is the "sacred space" of the liminal experience, where it is possible through ritually altered states (fasting, prayer, ritual intoxication, etc.) to communicate with beings at higher levels of consciousness. Most religious teachings counsel empathic communication, both with other humans and with divinity; to fail to communicate—in prayer, penance, and ritual activity—is to break faith with the community. In small communities the seamlessness of belief and daily activity ensures that such rarely occurs.

EVOLUTION OF SOCIOPOLITICAL COMPLEXITY

The renewal of state authority through ritual performance, on the other hand, is a more complicated matter. Annexation and conquest behoove states to encompass varieties of belief. To avoid the dangers alternative belief systems pose to the state, they must be deftly managed (e.g., through a syncretic state religion), severely discouraged, or eradicated.

AN APPLICATION: STATE HIERARCHY AND ENVIRONMENTAL CHANGE

Over a period of roughly 600 years, the Roman state vanquished its Celtic foes, moved from Republic to Empire, then steadily disintegrated, plunging all Western Europe into a Dark Age. This pivotal period in the history of the West offers a powerful illustration of the relation between climate and society and between organizational structure and long-term (*longue durée*) history.

Three major high-pressure systems characterize the climate of Western Europe: the Atlantic (Greenland high), the continental (Siberian high), and the Mediterranean (Azores high). Throughout Western Europe, from ca. 300 BC to 250 AD, the climate was Mediterranean—warm, dry, and unusually stable. The usually volatile climate of the continent was less variable than at any time since the middle Holocene. Climatologists refer to this period as the Roman climatic optimum (RCO) (Denton and Karlen 1973), when the Azores high-pressure system dominated the entirety of Western Europe. During this time, the relatively narrow zone of overlap (ecotone) between temperate northern climatic and biotic regimes (Atlantic–continental and semiarid Mediterranean) moved far to the north of its average twentieth-century position in southern France; vinyards flourished in England.

For Rome, the conquest of new lands and the dominance of a stable, warmer, and drier Mediterranean climatic regime over Western Europe created new sources and locales of wealth production; for example, new consumers and newly feasible growing regions expanded the wine and olive industries. This led to the expansion of certain forms of labor (e.g., slavery and debt-servitude), and to the imposition of more hierarchical forms of authority (Crumley 1987b, 1994; Clavel-Lêveque 1989). The climatic stability of the period meant that extreme weather events that usually characterized northwest Europe were absent, harvests were regular and abundant, and the growing alimentary needs of the population of Rome and other Imperial cities could be met. Industrial-scale farms, called *villae* and *latifundia*, grew specialized cash crops for urban markets. Entire regions (*provincia*) were consecrated to the production of specific goods for the huge urban market (Mommsen 1968).

For obvious reasons, the end of this halcyon period destabilized the Roman economy, but an equally critical factor was the diminished ability of hyperhierarchical Roman authority structure to adjust to less predictable circumstances (Crumley 1993, 1994, 1995b). We may now contrast the ability of rigid hierarchies (such as that of the Roman Imperial state) and more fluid heterarchical state organizations (such as were in place in climatically variable western Europe before the Roman Climatic Optimum (RCO) and persisted on the fringes of the Empire) to respond effectively to unforeseen conditions.

In all societies decisions are legitimated in many ways: birth, divine will, feats of valor, election, personal knowledge. In ranked societies, elites gain legitimacy and shape opinion by manipulating the potent instruments of religion, politics, and history; this is masked through public ritual, which demonstrates elite dedication to the collective well-being.

Personal ambition also plays an important legitimating role, as both exuberant self-advocacy and pious modesty may be combined, merging individual life goals with the long-term maintenance of the state. The reinterpretation of traditional cosmologies in favor of one's lineage or class is particularly effective, offering new readings that legitimate class hierarchy as a part of the natural world. Elite responsibility to society thus includes correct ritual practice, which ensures maintenance of the benevolent contract between deities and mortals, and the support of the arts, monumental construction, and other activities of public enrichment.

It is likely that climatic conditions worsened precipitously, not gradually, in the second century AD (Bryson and Murray 1977). There would have been increasing instances of crop failure (due to late spring frosts and/or cool, damp summers characteristic of the temperate European pattern) and ruin at harvest (hailstorms) or upon storage (blight).

Much was invested in the agricultural harvest in the European provinces: Of course, it fed the burgeoning population of Rome and other cities and it was a source of wealth for elites and financed their career aspirations. At a more visceral level, it mirrored the spiritual as well as the literal health of the Empire. The deities of the earth, harvest, and fertility (Tellus Mater, Juno, Hera, Ceres/Demeter, Artemis, Diana), along with those of weather and the elements (the sky, thunder) and war (Jupiter/Zeus, Athena), figured prominently in public worship (Vanderbroeck 1987; Wistrand 1979). The so-called Capitoline Triad is composed of Jupiter, Juno, and Minerva/Athena (the latter goddesses of the crafts, war, and wisdom). A temple dedicated to the triad (whose respective origins are pre-Greek, Etruscan, Minoan, perhaps even Indo-European) was begun in the first year of the Republic. It was located on one of the seven hills of Rome that served as both citadel and religious center.

EVOLUTION OF SOCIOPOLITICAL COMPLEXITY

Quite simply, it was inadmissible for Roman administrators of either the Republic or the Empire to fail to carry out their ritual duties to the triad, principal deities of weather and agrarian abundance. The success or failure of the harvest, however, was less critical during the Republic for two reasons. Not only would poor harvests have been uncommon during the RCO but because agricultural failure, like military defeat, would have been seen as the will of heaven, elite administrators would not have been held personally accountable.

The first two centuries of Imperial rule (mid-first century BC to mid-second century AD) were relatively prosperous, and the deified Emperors embodied divine will on earth. Beginning in the mid-second century AD, optimum climatic conditions began to deteriorate. The swift onset of agrarian collapse levied intricate economic and social costs and challenged sacred imperial authority. To account for widespread economic failure, Roman emperors were forced to reinterpret traditional religion or to embrace the mystical religious traditions of the eastern Mediterranean. Beginning with Aurelian (AD 215–275), who imported the Persian worship of the Unconquered Sun, the search culminated with Constantine's (AD 285–337) conversion to Christianity. Thus was fate safely transferred back into the hands of a deity, open to petition but possessing a decision-making logic that was unknowable and not subject to critique.

THE STATE OF THE STATE

Theories of self-organization and chaos give us a new, nonlinear way to think about human biological and cultural evolution and especially the formation of the state (Gumerman and Kohler 1994; Haken 1983; Harvey and Reed 1996; Kiel and Elliott 1996; Schieve and Allen 1982; Scott 1991). At each successive level of integration, new ordering principles come into play (*suspension of old forms*) (Jantsch 1982:348), drawing upon a store of knowledge (*preservation*) and providing creative solutions to new challenges (*transcendence of older forms*).

The examination of three key relationships has concerned us. The first, between self-organization and evolution, especially its stepwise character, may allow us to explain the relative suddenness with which states appear in the archaeological record. Self-organization also highlights the importance of communication, not for the convenience of decision makers alone but for the community as a whole. The emphasis of self-organization theory on holism serves as a reminder that the community, despite class and other divisions, functions as a collectivity. Service (1971b) correctly emphasizes the role of associations and organizations that cross-cut kin lines and that, while extant in most societies, profoundly characterize states.

The second relationship, between order and chaos, is the source of systemic creativity (that is, the potential of the system to change completely its parameters and become more richly networked). However, systems near chaos are subject to surprise. The human species and even individual human lives (an application of fractals) are all examples. The particular "surprise" examined here has been that of dramatic climatic change.

The third relationship is between hierarchy and heterarchy. While hierarchy undoubtedly characterizes power relations in some state societies, there are myriad coalitions, federations, democracies, and other examples of shared, counterpoised, heterarchical state power (see Chapter 10, this volume). All state systems have some heterarchical elements, just as all egalitarian societies have some relations that are hierarchical (e.g., age).

As sources of societal power diversify, markets expand, and belief systems and ethnicities multiply, more rigid hierarchies are unable to control diverse forms of social communication and thus to contain chaotic systemic behavior. The result is systemic collapse, whether through revolution or slow disintegration. Populist revolutions—those founded on democratic (heterarchical) ideals—do not necessarily reform elements of hierarchical state structure (Marx 1964).

Marked environmental change introduces surprise. As Gunn (1994) notes, the longer and more varied the history of acquired knowledge about a region's episodic climate, the more ably novel environmental conditions can be withstood. A long period of stability permits both the consolidation of power (hyperhierarchy) and the eventual loss of information about how to endure less salubrious times.

I have long maintained that states are relatively unstable, compared with other social formations. Such would be the case if states were to be considered complex dissipative systems. The flux of cultural energy and matter (that is, act and artifact) would vary over time, while state structure (the whirlpool) remained relatively stable. The research on strange attractors suggests the proposition that states have identifiable "basins of attraction," taking more hierarchical forms in environmentally stable periods, more heterarchical forms in periods of surprise. Democracies (states with both hierarchical and heterarchical elements) would be in theory the most stable of all, except that even then inequality of wealth, lack of cooperation, rigid ideology, and corruption introduce serious threats to stability (Midlarsky 1999). There are many directions from which surprise arrives.

ACKNOWLEDGMENTS

Thanks to William Marquardt for his insightful comments; to Julianne Maher, Rebecca Crist, and Jonathan Walz for calling my attention to related research; and to Joel Gunn for providing the chaos. This chapter is dedicated to the memory of my father, Howard M. Crumley.

Part *II*

The Emergence of Leaders

One of the most profound questions that arises in studying the evolution of cultural systems is why some people come to cede decision-making authority to other people: "Why would I let you make decisions for me and my family and then carry through on those decisions?" Obviously there are many possible answers to such a question—both historically at the level of individual actor/agents and structurally at the level of cultural systems. There have been diverse explanations offered for the emergence of centralized decision making that generally can be clustered into coercive and consensual kinds of models. In the coercive models, I do what you say because you have the means to make me comply. These means may be physically, economically, or even ideologically coercive. For consensual models, I do what you say because my compliance will result in some positive benefit to me and my family. Again these benefits may involve personal, psychological, or spiritual well-being.

Of course, there always has been some degree of centralized decision making in all human cultures. Parents make decisions for children, spouses make decisions for families, and elders commonly make decisions for kin groups. In the overall trajectory of cultural evolution, however, there was a profound change in social relations when centralized decision-making transcended the independence of autonomous kin groups and residential communities. This change is commonly recognized as corresponding to the emergence of chiefdom societies with the office of the chief having centralized decision making authority over a range of subordinate social and residential units. Ceding family and village autonomy to the leadership of chiefs, then, stands as one of the pivotal transformations in the evolution of culture. Examples of this transformation can be seen in virtually every world area, and while each individual case is historically unique, there are common patterns that run through them all.

The chapters in this section offer an introduction to the beginnings of political centralization and highlight the complex variables influencing the trajectory of cultures on the cusp of centralizing. Creamer bases her analysis on 10 years of field research in northern New Mexico, where she has been investigating the impact of European colonialism on the political organization of the indigenous Pueblo people. In Chapter 3, she opens a view into a complex system of large, independent villages of northern New Mexico, where incipient political centralization appears at the village level but village autonomy itself has not been transcended. Gilman has been studying the emergence of early centralized polities in southern Spain for almost 30 years. The complex at Los Millares represents what is probably the largest and most complex polity in all of Europe in the 3rd millenium B.C. His analysis in Chapter 4 represents a critical reassessment of a large body of work by himself and others, as he explores the rise and fall of incipient centralized societies in a challenging landscape. Creamer and Gilman together show that centralization is not an inexorable evolutionary outcome of demographic and economic growth and that societies can and do experiment with various manifestations of centralization without wholesale transformation of a village-based system to a permanent system of chiefdoms or states.

Chapter 3

The Origins of Centralization
Changing Features of Local and Regional Control during the Rio Grande Classic Period, AD 1325–1540

WINIFRED CREAMER

In the fall of 1540, the Spaniard Francisco Vasquez de Coronado with a following of soldiers, priests, and Indians entered what we know today as northern New Mexico. He was searching for gold and treasure, chasing the myth of the golden cities of Cibola. What he found instead was the vibrant culture of the Pueblo Indians. The Pueblos both impressed Coronado with the strength of their communities and to some extent confounded him because they did not fit the pattern of anything he was used to. Unlike the highly centralized and opulent civilizations of central Mexico the Pueblos did not have a king or even a dominant chief who could command the immediate obedience of his people. Coronado and subsequent Spaniards who explored northern New Mexico found to their considerable consternation that without a central Pueblo command structure, they really had little choice but to deal with each village independently, a cumbersome and costly way to both conquer and govern a new territory.

WINIFRED CREAMER • Department of Anthropology, Northern Illinois University, DeKalb Illinois 60115.

From Leaders to Rulers, edited by Jonathan Haas. Kluwer Academic/Plenum Publishers, New York, 2001.

In some ways, the organization of the Pueblos has continued to confound European explorers and visitors to northern New Mexico for the more than 450 years since Coronado's initial expedition, and this has been particularly true for anthropologists. One of the biggest questions for anthropology centers around the relationship among individual Pueblo communities. Today, each of the pueblos is politically, economically, and socially autonomous. Although there was certainly communication and interaction among the pueblos, they are all largely independent of each other, at times fiercely independent, as can be seen in negotiations over land or water rights and over casinos in the contemporary world. The presence of a relatively dense cluster of large, independent, and autonomous villages living side by side with each other is a unique phenomenon in the known ethnographic world. It is much more common to find that such village clusters are united in some form of interdependent tribal organization or arranged in a centralized and hierarchical chiefdom. But the Pueblos just do not fit either of these two traditional models of social organization.

In the discussion that follows, the development of leadership and hierarchy among the Pueblos is traced from the time of the earliest aggregated villages, around AD 1325 to the arrival of Europeans in AD 1540. This period of time has been called the Rio Grande Classic (Wendorf and Reed 1955). This analysis considers changes in demography (aggregation and population size), economic relations (production and trade), political change (e.g., warfare, the formation and maintenance of alliances), and the influence of religion and ritual. With these data, Pueblo society is shown to have been in a period of developmental transition and heading toward consolidation and coordination of communities. The long-term outcome might have resulted in centralized, hierarchical society, but this was never fully realized, as the process was interrupted by European contact in AD 1540.

DEMOGRAPHIC CHANGE

As would be expected with society based on corn cultivation, the archaeological record for the Southwest as a whole shows a fairly steady rise in population for 1000 years or more before the arrival of Europeans. At the same time, while we know that the population in general was growing, it has proved difficult to obtain specific information on population and demography for the northern Rio Grande region.

During our research on Pueblo settlement from 1450 to 1540, we identified 65 Pueblo villages in an area stretching from Isleta in the south to Taos in the north and from Pecos in the east to Jemez in the west. Some of these pueblos, such as Taos and Santo Domingo, are still occupied today, but the large majority were abandoned during the early colonial period. The smallest of these 65 sites include about 300 rooms and some sites include up to 3000 or more rooms. Together they represent what is probably the largest concentration of pre-European population centers anywhere north of the Valley of Mexico.

Population indicated by village size proves to be a deceptive gauge of total regional population, however. The 65 sites recorded in northern New Mexico are big and obvious. During the latter half of the Rio Grande Classic, the overwhelming majority of the people in this region were living in one or another of these large pueblos. Therefore, we know where almost everyone was living, but we do not know how many people were actually living in the area at any one time.

It is difficult to determine how many of these sites were occupied in 1500, for example, and how many people were living in each of the occupied villages. The dates we have for these sites generally overlap, so most if not all could have been occupied during the late 1400s. But we do not know if there were actually people living in all villages at that time or how many people may have been living in any particular village.

Part of the problem is that these villages all had different life spans. When we tested a sample of 13 of these sites, we found that some of them had very shallow and sparse trash deposits, while others have thick heavy deposits 2m deep (Creamer 1996; Haas and Creamer 1998).

A more serious and unexpected problem with estimating population dynamics is that the large, ancient Pueblo villages tend to have complex histories of what can be called "sequential occupation." For example, Pueblo San Marcos in the Galisteo Basin had a total of about 3000 rooms and was occupied for a period of more than 300 years from the 1300s into the colonial period. Although there are complex formulas for figuring out the population of ancient villages based on floor space, number of hearths, or numbers of broken pots per decade, they generally come down to an average of about one person per room in the Southwest (Lofgren and Turner 1966). Applying that very general formula to San Marcos yields an estimated population of approximately 3000 people.

Unfortunately, this formula does not really tell us about population dynamics. The pueblo could have grown gradually to 3000 people and then declined gradually until abandonment. Or, it could have grown rapidly to 3000, stayed at or close to that maximum for most of its 300 years and then undergone rapid depopulation. Testing at San Marcos indicates a quite different picture, however. Based on intensive surface collections and subsurface test excavations, it appears that large blocks of rooms were occupied at different times. Thus, one part of the site was occupied in the 1300s and then abandoned. Later, in the fifteenth and sixteenth centuries, other roomblocks were occupied and abandoned. In the seventeenth century yet another set of roomblocks was built in an area in the immediate vicinity of the Catholic mission. Work conducted at other large, contemporaneous pueblos in the northern Rio Grande has shown a similar pattern of sequential occupation and abandonment of a specific site locale (e.g., Creamer 1993; Lambert 1954). Though we know there was aggregation, sequential construction and use indicate a much smaller maximum site population than the "1 room = 1 person" formula suggests. Our more intensive research at Pueblo Blanco in the Galisteo Basin specifically suggests that from 30% to 50% of the village was occupied at any given time (Creamer 1996; Haas and Creamer 1998).

This pattern of sequential occupation alters our image of the timeless, unchanging pueblo that has emerged in the more recent historic era. The latter image illustrates the significant impact of Spanish colonialism on the pattern that prevailed in pre-European days. It appears that different factors influenced Pueblo settlement patterns both before and after the intrusion of colonists into the area. Before the arrival of colonists, the pattern of sequential occupation and abandonment seems to have been primarily a function of two variables: availability of wood for fires and building and more importantly arable land. When Spanish explorers arrived, they described pueblos as extremely large, multistoried, fortified localities, apparently reflecting the fear they felt. The reality of these extensive villages may not have been as fearsome as the explorers' first impressions.

Sequential occupation was a likely adaptive strategy for Pueblo people of the fifteenth century. They picked an area with reasonable water, arable lands, and nearby wood sources, built houses, and lived there for 30, 50, maybe even 75 years. At the point when nearby fields were no longer productive and wood for fires and remodeling was an impractical distance, it became expedient to move to a new location with fresh fields and sources of wood; this was the dominant practice across the prehistoric Southwest for more than a millennium (Plog 1997). Since arable lands and local forests will replenish themselves if allowed to lie undisturbed for a long stretch, the question is really whether fields lay fallow while others were used or whether whole sites were abandoned between construction episodes.

This strategy of sequential site occupation was significantly altered by the arrival of Spanish colonists. First, the early Europeans introduced new domesticated plants and animals into the local economy which had a profound effect on pueblo subsistence. New crops allowed for more efficient crop rotation and exploitation of more land for agriculture. Domesticated animals not only added important sources of protein to the diet but also provided fertilizers to greatly prolong the life of local field systems. This reduced the impetus for moving villages every 20 to 30 years. Village mobility also was impacted by the Spanish colonial administration of the region. One of the first steps of the colonial governors was to limit the movements of the Indians by encouraging and forcing them to live in proximity to Catholic churches and *visitas*. The immigration of Spanish colonists into the area further restricted the movement of Pueblo peoples. Settlers coming into the area quite reasonably (from their perspective) would have put down their farmsteads in areas that were not immediately occupied and farmed by the Indians. These immigrant colonists then had the effect of occupying many of the open spaces that the Pueblos had been using for "swing space" in their pattern of sequential occupation. Thus, the pattern of large, stable pueblos that is so well documented in the historic and ethnographic records of northern New Mexico is ultimately an artifact of the interaction experience between the Indians and the Spanish colonizers.

The sequential occupation of these large sites carries two significant implications in terms of the organization of polities on a regional level. First, the regional population and the population of individual sites were much lower than can be inferred from surface surveys and historic excavations from earlier in the twentieth century. Thus, rather than 65 or more sites occupied contemporaneously in the fifteenth and sixteenth centuries, there were probably never more than about 30 to 40 of these large sites occupied at any given time. Instead of a regional population of 60,000 to 100,000 people, the total size of the population over the 17,000-km^2 area may have been occupied by something more on the order of 20,000 to 30,000 people. The difference in both density and absolute number of people in the area means that there was less pressure and opportunity for political interaction and organization beyond the level of the village.

The second implication of the pattern of sequential occupation is that the number of people living in a village at any given time was frequently much lower than would be indicated by the room count. Although there may well have been a few precontact villages with more than 1000 residents, they were more the exception than the rule. Lower community population figures would mitigate the need for centralized village administration.

ECONOMIC CHANGE

During the Rio Grande Classic period there was increased economic differentiation within the region that is manifested in changing ceramic style, and evidence of trade in ceramics and lithics. Habicht-Mauche has explored economic relations of northern Rio Grande ceramic production, concluding that village specialization developed during the 13th through the 15th centuries (1995, Snow 1981).

Thirteenth Century

Stylistic similarities in black-on-white ceramics across the region during the thirteenth century indicate regular contacts and some regionwide interaction (Habicht-Mauche 1995:174, Figure 7.4). The boundaries of the region at this time can be defined by the distribution of Santa Fe black-on-white ceramics. This direct association of stylistic attributes with the boundaries of a regional system assumes regular communication among potters, without restriction on use of resources by any individual or group. Household production of goods without specialization created weak bonds of cooperation that seem to have held individual villages together in a loose internal network. Ceramic exchange appears to have been extensive, yet regional political interaction other than exchange was ephemeral. Sites were small and there were no central places, evidence of only low levels of warfare and no signs of an imbalance of exchange.

Fourteenth Century

Early in the fourteenth century, an increase in variability in black-on-white pottery has been identified. From north to south the new types included: Talpa/Vadito B/W produced in the Taos area, Wiyo B/W produced north of Santa Fe, Galisteo B/W produced south of Santa Fe, Rowe B/W produced to the east, and Jemez B/W produced in the west in the Jemez Mountains (Habicht-Mauche 1995:176, Figure 7.6). These ceramic types vary in style, but are very similar in form and presumably function. Habicht-Mauche suggests that increase in population fostered a level of competition and conflict that led to boundary formation. Alliances were based on exchange and were marked by ceramic style (Habicht-Mauche 1995:192–193). These newly formed boundaries appear to have been weak and highly permeable in terms of the movement of resources by exchange.

In the latter part of the fourteenth century, we see the introduction of distinctive glaze wares and biscuit wares to replace four of the preceding black-on-white types. Only in the Jemez area did ceramics stay largely unchanged from the fourteenth through the seventeenth centuries. Thus, it appears that by the end of the fourteenth century the northern Rio Grande was forming into three socioeconomic units differentiated by ceramics and networks of communication and interaction.

Fifteenth Century

By the beginning of the fifteenth century, ceramics decorated with a unique lead-bearing glaze were being made in the southeastern third of the northern Rio Grande region. The glaze wares and the glaze ware subregion have been intensively studied and accordingly provide the best-defined picture of economic interaction. Two models of production and distribution of the glaze wares have been proposed. Anna Shepard's (1942) classic petrographic studies of New Mexican glaze wares identified multiple centers of production. Her work implied that there was localized production of ceramics in several different zones with varying locally available resources (Figure 3.1) (Shepard 1942). Lead would have been obtained either directly by villages around the restricted sources in the Ortiz Mountains or through trade with villages near those sources. Snow's (1981) analysis of protohistoric exchange supports Shepard's model of localized production by concluding that family-level production could account for the patterning in material culture that has been recorded.

A contrasting model has been offered by Helene Warren (1969), who proposed more centralized production of glaze wares. She argued that San Marcos Pueblo was a major production center during the fifteenth century, and that this production center shifted to Tonque Pueblo during the sixteenth century based on temper analysis (Warren 1969, 1979) (Figure 3.2). She assumed lead ore was controlled by the producer villages, and the temporal shift from one production center to another reflected a shift in political power.

Subsequent work has tended to support a model of more localized production. Recent X-Ray Flourescence (XRF) analysis of glaze ware ceramics from six protohistoric sites in the Galisteo Basin and one further west (Kuapa LA 3444) suggests that San Marcos Pueblo was not the center of glaze production in the Galisteo Basin and other villages, probably every one, produced glaze wares (L. W. Reed 1990:146). Analysis of the variability of glaze varieties among sites in the Galisteo Basin tends to confirm the localized production of ceramics at each of the villages (Creamer et al. 2000). Thus, for example, ceramics identifiably from Tonque (based on temper) did not make up a significant proportion of the late glaze ceramics anywhere in the Galisteo Basin. These analyses support Shepard's earlier indication of multiple production centers, rather than Warren's model of focused production. While there may have been specialist producers within each village, there is no firm evidence of village-wide specialization.

Another ceramic type, biscuit ware, replaced black-on-white ware produced in the northern part of the Rio Grande region beginning about AD 1360 (Schaafsma 1995). Analysis has indicated some degree of standardization, with an implication of centralized production (Hagstrum 1985). While facilities for standardized production, such as workshops and formally constructed kilns, have not been recovered archaeologically, a trend toward subregional specialization in ceramic production is clear by the fifteenth century.

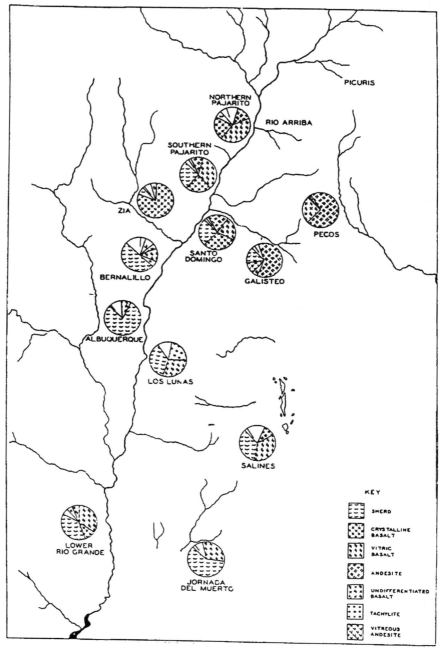

Figure 3.1. Localized production of glaze-decorated ceramics as indicated by Shepard (1942).

ORIGINS OF CENTRALIZATION 45

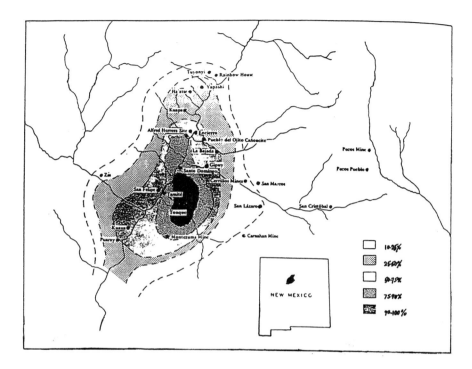

Figure 3.2. Specialized production of glaze-decorated ceramics as indicated by Warren (1974).

The distribution of distinctive decorated ceramics during the fifteenth century illustrates the ubiquity of ceramic exchange, as sites in each subregion have decorated ceramics from each of the other subregions (Table 3.1). Quantities, however, suggest varying levels of exchange. In the biscuit ware subregion, very few fragments of either glaze ware or Jemez black-on-white were found, and at least 98% of the decorated ceramics at each site tested were the locally made biscuit wares. Ceramic distribution within the glaze ware subregion differs only slightly. More than 90% of the decorated ceramics at each site were glaze wares. Biscuit ware comprised less than 10% of decorated ceramics, while Jemez black-on-white was present in only trace amounts. Sites in the Jemez subregion are more variable. Only one sherd of biscuit ware was recovered among the four sites tested. Glaze ware was also rare at one site, where 99.5% of decorated ceramics were Jemez black-on-white. At the three other sites tested in the Jemez subregion, however, glaze ware comprised from 13 to 27% of decorated ceramics.

These data suggest that subregional boundaries were becoming stronger, limiting exchange between villages within the biscuit- and glaze-pro-

Table 3.1. Proportion of Decorated Ceramic Types

	Biscuit		Glaze		Jemez black-on-white	
	No.	Percent	No.	Persent	No.	Percent
Biscuit ware sites						
Hupobi (LA 380)	512	98.0%	3	0.5%	8	1.5%
Ponsipa (LA 274)	694	98.7	5	0.7	4	0.6
Pose'ouinge (LA 632)	861	99.3	5	0.6	1	0.1
Poshu (LA 297)	557	99.8	1	0.2	0	0.0
Sapawe (LA 306)	1956	99.5	9	0.5	0	0.0
Glaze ware sites						
Kuapa (LA 3444)	43	8.1%	483	91.5%	2	0.4%
Pueblo Blanco (LA 40)	19	5.8	306	93.6	2	0.6
San Marcos (LA 98)	18	1.8	1002	98.1	1	0.1
Jemez black-on-white						
Kiatsukwa (LA 132/133)	0	0.0%	1	0.5%	181	99.5%
Kwastiyukwa (LA 482)	0	0.0	69	18.0	316	82.0
Nanishagi (LA 541)	1	0.5	25	12.9	168	86.0
Seshukwa (LA 303)	0	0.0	89	27.4	236	72.6

ducing subregions and between the biscuit- and Jemez black-and-white-producing subregions. However, there appears to have been substantial distribution of glaze ware in the Jemez subregion, though the exchange of ceramics was apparently one way, with quantities of glaze ware going to the Jemez subregion but little or no Jemez pottery entering the glaze ware sites. It seems likely that the glaze wares were being exchanged for nonceramic goods or services. Further, there may have been greater political connection between glaze-producing and Jemez sites, resulting in a less clearly defined economic boundary between the two.

The distribution of nonceramic artifacts adds insight into economic organization and interaction in the region. Obsidian, for example, was obtained from only four primary sources in the Jemez Mountains (Baugh and Nelson 1987). Though obsidian tools and flakes from these sources are found at every fifteenth-century village, quantities vary sharply from place to place (Table 3.2). Interestingly, there were significant differences in the distribution of obsidian between sites in the Jemez Mountains. At two of the Jemez sites tested, Nanishagi and Seshukwa, 60 to 70% of the chipped stone was obsidian. However, Kiatsukwa and Kwastiyukwa, two other sites tested in the Jemez region had only 14–15% obsidian. This may distinguish specialized quarrying or tool producing villages from other sites. At Kuapa, the glaze ware site closest to the Jemez region, 26% of the chipped stone was

ORIGINS OF CENTRALIZATION 47

obsidian, while more distant glaze sites in the Galisteo Basin had from 1 to 7%. Compared with ceramic distribution, this suggests that some but not all villages in the Jemez Mountains were supplying obsidian to those in the Galisteo Basin in exchange for glaze-decorated ceramics and/or the contents of such vessels, with Kuapa perhaps acting as middleman. In contrast, villages in the biscuit ware subregion had little obsidian, less than 1% of total chipped stone in each of the five sites tested (Futrell 1998). This supports the idea that the boundary between the Jemez and biscuit subregions was a barrier to exchange. At Galisteo Basin sites, most stone tools were made from fine-grained dark basalt, but all have small points and scrapers made of Pedernal chert, a source in the biscuit ware area.

Other materials were distributed throughout the northern Rio Grande. From the Plains, bison products reached the northern Rio Grande and from the Salinas Pueblos, salt. Turquoise and fibrolite axes (Snow 1973, 1981; Weigand et al. 1977) also were exchanged across the region. Small amounts of turquoise, mostly coming from the Cerrillos Hills in the glaze-producing Galisteo Basin, have been recovered from protohistoric sites throughout the northern Rio Grande region. Cerrillos turquoise is found at all Galisteo Basin sites and is quite common at the pueblo of San Marcos, which is closest to the Cerrillos Hills. This much higher frequency at San Marcos may be one of the few signs of centralized control over the procurement of a single

Table 3.2. Proportion of Obsidian in the Lithic Assemblage of Each Site

Site	Amount of obsidian (% of lithic assemblege)
Biscuit sites	
Hupobi (LA 380)	0.6%
Ponsipa (LA 274)	0.4%
Pose'ouinge (LA 632)	0.8%
Poshu (LA 297)	0.7%
Sapawe (LA 306)	0.4%
Glaze sites	
Kuapa (LA 3444)	26.2%
Pueblo Blanco (LA 40)	0.9%
San Marcos (LA 98)	7.1%
Jemez sites	
Kiatsukwa (LA 132/133)	14.2%
Kwastiyukwa (LA 482)	15.2%
Nanishagi (LA 541)	70.9%
Seshukwa (LA 303)	62.7%

resource. However, San Marcos is located several kilometers from the Cerrillos Hills source, and thus does not appear to have exercised physical control over the turquoise source.

Overall, the distribution of decorated ceramics and other materials do not indicate marked levels of either specialization or centralized decision making in the production or distribution of economic resources.

AGGREGATION

If we look at the northern Rio Grande prior to 1300, we find pueblos of 15 to 30 rooms were common, while those of 100–200 rooms such as Pot Creek (Wetherington 1968) and Pindi (Stubbs and Stallings 1953) were rare. People lived in a range of communities from hamlets of 2 or 3 families up to small villages of 10 to 20 families (Cordell 1979; Stuart and Gauthier 1984). Major change occurred during the fourteenth century with growth in the size and numbers of pueblo villages in the northern Rio Grande. Immigration from the north and west was sustained by adequate rainfall and proximity to river and stream bottomland. The aggregation that began at this time continued until European contact. Villages got larger, and larger until by 1500 virtually everyone was living in pueblos of more than 300 rooms.

Aggregation on this scale suggests that over time there was need for increasing coordination of water use, land use, and hunting. People also needed dispute resolution, mediation of contacts with other villages, a mechanism for making decisions that affected the group as a whole, and ways to foster village solidarity. The introduction of the enclosed plaza village layout has been suggested as evidence that religion, specifically the Pueblo katsina cult, was the vehicle for both group decision making and ritual to promote village unity (Adams 1989). Control of ritual knowledge through membership in secret societies was an additional step toward hierarchy (Brandt 1994). Competition among villages for farmland and hunting and collecting territory increased to the extent that social boundaries became more evident during the 1400s (Hunter-Anderson 1979).

Paul Reed (1990) has analyzed the spatial clustering of pre-European villages in the northern Rio Grande for indications of political alliance by examining large (> 400 room) villages occupied during the Rio Grande Classic period (AD 1325–1540). Using the nearest-neighbor statistic, he assessed six groups of sites and found four of these to be significantly clustered. The groupings include sites in the Jemez Mountains and others in the Chama/Ojo Caliente drainage, the Galisteo Basin, and the Albuquerque area (Figure 3.3).

ORIGINS OF CENTRALIZATION

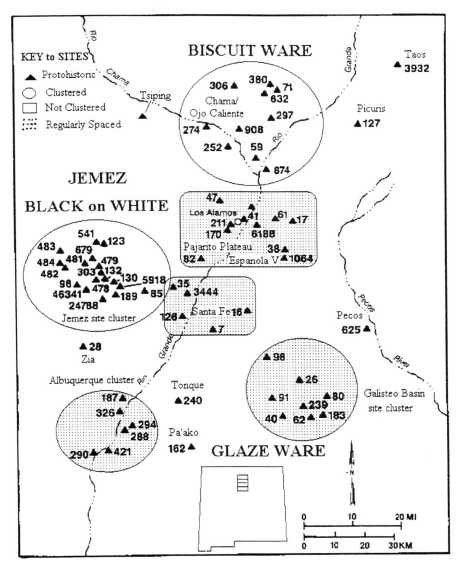

Figure 3.3. Village clusters and isolated villages (after P. Reed 1992).

Additional testing of these groups using the linear nearest-neighbor statistic indicated that within the groups two areas not viewed as significantly clustered exhibited regular spacing of sites (the Santa Fe and Pajarito Plateau). Two of the clusters—the Galisteo Basin and the Albuquerque area—

also were regularly spaced. The Jemez Mountains, and the Chama/Ojo Caliente (biscuit ware) zones were more tightly clustered. Reed (1990) suggests that regular spacing is an indicator of competition among villages, while Hunter-Anderson (1979) suggests that the aggregated village itself was a response to competition and a need for intensification of production.

Other villages in the northern Rio Grande, however, are not part of any group or cluster. Taos and Picuris, for example, at the north end of the region are each isolated, single villages. Adjacent to the Galisteo Basin but outside the cluster of villages is Pecos, which was reported as an independent center by the Spanish in 1540. Further south is Pa'ako, located on the east edge of the Sandia Mountains, and Tonque, north of Albuquerque, apparently astride an arroyo long used as an east–west thoroughfare (Warren 1969). On the west side of the region is Zia Pueblo, located in between the Albuquerque and Jemez Mountains site clusters.

The implication of site patterning is that there were varying degrees of autonomy and integration across the region. Some villages, physically isolated, appear to have been more socially and politically autonomous than others. These isolated villages, such as Tonque, Zia, or Kuapa, have specific features or attributes in architecture, material culture or site layout that make them distinctly different from other sites in the same general vicinity. In contrast, villages in the more tightly clustered groups—the Jemez Mountains, and Chama/Ojo Caliente—were much more homogeneous. All sites are constructed in the same masonry style in the Jemez subregion, while sites in the Chama/Ojo Caliente zone all have remarkably similar layouts of extensive adobe room blocks surrounding wide-open plaza areas. The sites also are similar in terms of general site location, the situation of kivas, and artifact frequencies.

The grouping of sites in the northern Rio Grande that exhibits regular spacing among member sites illustrates conflicting as well as unifying forces. The Galisteo Basin cluster shows significant homogeneity in ceramic production. The principal resource—lead for glaze-decorated ceramics—was central to them all. At the same time, variability in architecture, lithics, and even techniques of cultivation in the Galisteo Basin (Lightfoot 1993) suggests the occupants of each village were each pursuing independent economic and social strategies.

At Pueblo Blanco, the earliest rooms at the site were all adobe with stone foundations. Subsequent construction was a combination of adobe and stone. Some walls were solid adobe, some solid stone, and others a mix of the two materials. At San Cristobal, not far away, the early occupation of the site is all adobe, while a later occupation is completely constructed of stone. San Marcos is mostly adobe throughout its occupation, but with some stone used in most room blocks. Pueblo Largo, in contrast, is constructed

ORIGINS OF CENTRALIZATION 51

completely of stone masonry. It also is interesting to note that at Arroyo Hondo, just to the north of the Galisteo Basin, one of the earliest constructions at the site was all of stone, while later ones were all abobe, the opposite pattern of what is found in most Galisteo Basin sites.

WARFARE

The large aggregated pueblos that resulted over time seem natural to us today, although in many ways pueblos are a troublesome settlement type that demands explanation. From an economic standpoint, farmers in a preindustrial agricultural economy need to be close to their fields. One has only to look at the vast cornfields of the American Midwest to see the pattern of farmers building their houses right in the middle of their fields. And indeed, for the first 1000 years or so of corn agriculture in the Southwest there was a pattern in the archaeological record of small hamlets spread across the landscape putting local farmers in direct proximity to their fields. What was it then that changed in the fourteenth and fifteenth centuries to cause farmers to abandon the convenience of being near their fields and aggregate their families together in large villages? It appears that the isolated family hamlet became vulnerable to attack from raiding parties searching for corn and other resources.

There are a number of other kinds of evidence of conflict and war among the Pueblos (see Haas and Creamer 1996), but the physical location and layout of the pueblos themselves provide the clearest clues indicating that peace did not always prevail between the Pueblo communities prior to the arrival of the first Europeans. The most common location of large sites in the Jemez region, for example, is at the tip of long, narrow, steep-sided mesas in strongly defensive localities. Tsiping, an isolated site outside the Chama/Ojo Caliente site cluster, was built on a mesa top. The village was then surrounded by walls, forming an impressive fortress.

In the Galisteo Basin, Pueblo Largo covers a hilltop overlooking the valley floor. (It also is significant to note that many of the rooms at Pueblo Largo were burned in an intense fire at abandonment.)

Beyond these fairly extreme defensive positions, all of the northern Rio Grande pueblo sites exhibit defensive features, however. For example, there are very few ground floor doorways in any of these sites prior to the arrival of Europeans. Ground floor rooms were entered by way of ladders through roof top entries. Villages were sometimes a solid mass of rooms, as at Taos Pueblo (Figure 3.4a). A few other villages consisted of unconnected room blocks (Figure 3.4b). More often, sites consisted of blocks of 20 to 150 rooms, in a three-sided or quadrilateral arrangement around an open area

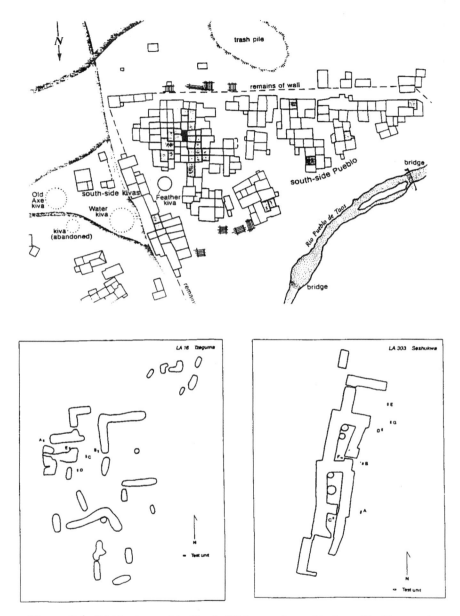

Figures 3.4. (A) Massive pueblo (Taos). (B) Scattered roomblock pueblo (Tzeguma). (C) Open plaza pueblo (Seshukwa).

or plaza that sometimes included a kiva (Figure 3.4c). Narrow openings at the ends of room blocks provided the only access to enclosed plazas. This restricted access to these areas; only a few people at a time could enter inner, enclosed plazas. A large village would have required planning to make sure that construction was consistent, especially if defense was the goal.

The changing use of plaza areas over time also has defensive overtones. Beginning with Arroyo Hondo we find sequential components of occupation dated to 1300–1350 (component I) and 1360–1420 (component II) (Creamer 1993). Plaza areas included a variety of features indicating outdoor activities. Deposition was shallow, with a hard-packed central area and less heavily compacted edges. The small quantity of trash recovered from plaza areas suggested that these public areas were purposely swept and kept clean. Work areas on rooftops were used in the same way that plaza work areas were used, with excavation data indicating a shift toward use of rooftops over plazas during Component 2 (Table 3.3).

This combination of rooftop and plaza work areas appears to have changed during the century before the arrival of Europeans. Testing at 13 large villages occupied during the fifteenth century revealed that plazas most often were the location of primary trash disposal. Plaza areas at many sites had deep trash fill, up to 2m along the margins of plazas (Creamer 1991). Of the 13 sites tested, nine had trash deposits in plaza areas only, three had trash deposits in both plazas and outside the periphery of the roomblocks, and at one site trash areas were exclusively outside the boundaries of the roomblocks (Table 3.4).

The trend toward preferring rooftops for activities may have originated with an increasing concern with defense. Rooftop work areas offered outdoor work space similar to plazas but with somewhat more difficult access. Pedro de Castaneda, chronicler of the Coronado expedition, noticed the Pueblos concern with defense during the first European exploration of the Rio Grande region:

> The houses have no doors on the ground floor. The inhabitants use movable ladders to climb to the corridors, which are on the inner side of the pueblos. They enter them that way, as the doors of the houses open into the corridors on this terrace. The corridors are used as streets. The houses facing the open

Table 3.3. Work Areas at Arroyo Hondo

Component	Rooftop Work areas	Percent excavated rooms with roof work area	Rooms with roof mealing bin	Plaza mealing areas
1	20	20%	18 (18%)	6
2	12	23%	12 (23%)	6

Table 3.4. Location of Trash Middens at Pueblo Villages.

Site name	Construction	No. of Rooms	Trash in Plaza	Trash outside Plaza
Galisteo Basin				
Pueblo Blanco	Masonry and adobe	1450	In	
San Marcos	Adobe	2500	In	
Jemez Mountains				
Kiatsukwa	Masonry	500	In	
Kwastiyukwa	Masonry	1250	In	Out
Nanishagi	Masonry	450	In	Out
Seshukwa	Masonry	1100		Out
Rio Grande Valley South of Santa Fe				
Kuapa	Masonry	450	In	
Tzeguma	Adobe	500	In	
Chama-Ojo Caliente Drainage				
Hupobi	Adobe	900	In	
Ponsipa	Adobe	250		In
Poshu	Adobe	1810	In	Out
Posi	Adobe	750	In	
Sapawe	Adobe	2000	In	

country are back to back with those on the patio, and in time of war they are entered through the interior ones. The pueblo is surrounded by a low stone wall. Inside there is a water spring, which can be diverted from them. The people of this town pride themselves because no one has been able to subjugate them, while they dominate the pueblos as they wish. (Hammond and Rey 1940:256–257)

Conflict between villages also can be seen in various oral histories of the pueblos themselves. The following is from a recent history of the pueblo of Santa Ana:

Kwiiste Puu Tamaya [Santa Ana], the tradition says, was blessed with flourishing fields of corn and squash. The women of the Tsiiyame [Zia] village would cross the river each day to help the Tamayame grind corn. But the stories recall that the men of the Tsiiyame village soon grew jealous because their women were crossing the river, and in their anger, they began to plot against the Tamayame. . . . When the people of Kwiiste Puu Tamaya learned of the plot, they decided to move on. They gathered food and supplies for the journey, packed all that they could carry, and then set fire to the village. Carrying their belongings, they moved south until they came to Kene'ewa (San Felipe Mesa), just southeast of Siiku (Mesa Prieta) and east of the Rio Puerco. On top of the mesa, they sought shelter from their enemies. Later, as the threat of attack became more remote, the Tamayame began to plant corn in the wide valley at the base of Siiku. (Bayer 1994:7)

ORIGINS OF CENTRALIZATION 55

The continuing focus on defense illustrates the intense competition among pueblos. The aggregated village of enclosed plazas was itself a successful adaptation from the range of earlier forms. The fact that villages were rebuilt over and over in a single location, while varied land use strategies were employed from that permanent base further suggests there was not always enough land available for cultivation to sustain the resident population. By the time Europeans arrived in 1540, conflict had not yet risen to the level seen in many chiefdoms, such as those in Central America or Oceania. A war cult and war leaders were part of pueblo life (Ellis 1951), but the transformation to centralized command over fighters had not yet occurred.[1]

RITUAL AND RELIGION

Until recently, most archaeological interpretations of pueblo political organization focused on determining whether the ancient Pueblos represented "tribal," "chiefdom," "state," or some other established level of societal complexity (Johnson 1989; Spielmann 1994; Upham 1982). Dissatisfaction with the categorization of large Pueblo villages as "tribes" (Creamer and Haas 1985) or as "complex tribes" (Habicht-Mauche et al. 1987) indicates a need to consider how concepts of leadership, power, and political organization may apply to Pueblo society (cf. Neitzel 1999). During the Classic period, what have been called "polities" were either forming or already in place in the northern Rio Grande. "Polity" may currently be a more acceptable term than "chiefdom" or "state" because it is less specific and can encompass the very "statelike" polities described by Wilcox (1991) or the very "tribelike" polities described thus far here.

A key to understanding the development of complexity in the northern Rio Grande may be the role of ritual in society. Religion and ritual played a substantial role in Pueblo life. The status of ritual practitioners was high. Archaeologically, there is little evidence of status differences among dwellings in Pueblo structures, but clear evidence of rooms devoted to ritual meetings or storage of objects. Ethnography of Pueblo society emphasizes the importance of ritual in daily life but characterizes political leadership as weak (Benedict 1934:100–103).

For example, Brandt (1994) describes the importance of secrecy to Pueblo religion and a connection between possession of ritual knowledge and personal power. Brandt shows the use of secrecy to reserve knowledge to a small group, often a religious society. Others have shown an associa-

---

[1]The Pueblo Revolt, which took place in 1680 and unified most of the Rio Grande pueblos, shows that centralization was well within the realm of possibility.

tion between religious rank and clan membership and between clan and the distribution of land (Levy 1992). In this way, the ritual system may have masked the identity of decision makers who instituted and maintained inequalities in resource distribution.

The relationship of secrecy and the possession of secret knowledge to power, has also been discussed by Helms (1979) in her ethnohistoric study of sixteenth-century Panamanian chiefdoms. These cross-cultural data ultimately show how individuals with secret knowledge are in a privileged position to lead critical rituals associated with rainfall and fertility. It is this control of knowledge, whether in the katsina cult or the *quevi* of ancient Panama, that plays a key role in converting religious leadership into a more stable and powerful combination of sacred and secular rule.

Among the Rio Grande Pueblos, increasing power of religious leaders is evident in the introduction of Great Kivas around AD 1300. This provided the opportunity for ritual to include inhabitants of more than one village, a form of alliance. The shift from constructing Great Kivas to using enclosed plaza areas within each pueblo for ritual practice (Adams 1991) suggests a further accumulation of power by priests, elders, or other ritual practitioners. Construction of a Great Kiva took effort by a sizable group, but construction of an enclosed plaza took the cooperation of the entire village, a clear extension of priestly influence. Early chroniclers took little note of Pueblo religion other than to condemn the masked dances that were central to public ritual. As a result, Pueblo religion rapidly disappeared from the sight of Europeans, though we know it survived within the pueblos. The little we know of leadership within villages from historic sources comes from accounts that describe leaders as elders who could convene meetings to discuss issues but whose power did not extend beyond one village (Hammond and Rey 1966). Ethnography suggests religious leadership was paramount in the pre-European era (e.g. Bayer 1994:55). Taken together, it seems that early sixteenth-century Rio Grande pueblos were led by one or more religious practitioners, aided by the counsel of a group of elders.

CONCLUSIONS

At the time Europeans arrived in the Rio Grande Valley in 1540, competition among villages was prominent and separate groups or clusters of villages were emerging. No single strategy of alliance, confederacy, collaboration, hierarchy, heterarchy, or warfare had yet proved to be more successful or effective than any other. The variation observable archaeologically in settlement pattern, ceramic style, and exchange network reflect experimentation among political strategies in the context of the regionwide competitive environment. The northern Rio Grande region was a system in political transition.

Villages were led by members of Pueblo "secret societies" or other religious practicioners. Alliances appear to have been growing among the site clusters in the Galisteo Basin pueblos, the Chama/Ojo Caliente, Jemez Mountains, and Albuquerque subregions. Accounts written of Coronado's expedition to the region name "provinces" that appear to be site clusters or groups of villages speaking a common language (Habicht-Mauche 1988). The bonds holding these alliances together were weak, indicated by the fact that the Spanish found they could not provision themselves by requesting supplies from the leader of one of the villages in the region:

> ... he sent for an Indian chief of Tiguex with whom we were already acquainted and with whom we were on good terms [Juan Aleman].... The general spoke with him, asking him to furnish 300 or more pieces of clothing which he needed to distribute to his men. He replied that it was not in his power to do this, but in that of the governors'; that they had to discuss the matter among the pueblos; and that the Spaniards had to ask this individually from each pueblo. (Hammond and Rey 1940:224)

Feinman (1998) has suggested that Pueblo political organization was fundamentally corporate in nature, relying on group action toward common interests rather than on individual leadership. This view helps explain the European accounts of numerous local leaders in the region but no paramount chief. Further, pressures for group cohesion also could explain the similarity in housing and in mortuary customs within Pueblo villages and among pueblo sites in the northern Rio Grande. However, the fact that leadership may have been corporate in nature does not mean that the Rio Grande was a region with highly complex political organization. Relatively low population density and the difficulty of generating surplus in an arid region having unpredictable precipitation suggest centralized polities or strong alliances were difficult to achieve.

The alliances that had been developing among Rio Grande pueblos could not withstand even the initial visit of Europeans. Apparently there was virtually no surplus in the exchange system. Coronado's winter 1541 encampment depleted clothing, food, and housing stock to such an extent that the area was not reoccupied: "However, the 12 pueblos of Tiguex were never resettled as long as the army remained in the region, no matter what assurances were given them" (Hammond and Rey 1940:233–234).

By the time the first permanent colonists settled in northern New Mexico in 1598, there appears to have been only vestiges of the alliances and power differentials among pueblos that appeared to have been heading toward more centralized polities during the early sixteenth century. A brief period of regional unity in the late 1670s resulted in the only successful Indian revolt in the New World in 1680. This confederation occurred after 50 years of conflict between European authorities of church and state over how best to employ the new Puebloan members of the European realm. The revolt was short-lived, however, and by the turn of the eighteenth century, Pueblos were atomized tribal groups in the autonomous village model seen today. Colonization truncated political change, but the region does provide some potentially important insights into the kinds of evolutionary processes that appear to be the immediate antecedents of political centralization. The archaeological record for the period of aggregation and political diversification in the northern Rio Grande lasted only for about 150 years, less than the measurable time increments archaeologists commonly apply in most other world areas. Thus, from an archaeological perspective, the period of political experimentation was very short and should be viewed as a period of evolutionary transition rather than as a stable adaptive system.

The transition from leader to ruler involves the transformation of society. A leader acts as head of a group based on knowledge, skill, and experience. The group ranges in size from a family to a village or group of villages. A ruler's position is based on rank, ancestry, knowledge restricted from others, and training. Rulers are likely to govern more than one locality, living near features of centralization, such as markets, and storage structures, as they need to control resources. A society moves from requiring that leaders demonstrate their fitness to accepting rulers based on some long-forgotten (or mythical) fitness of an ancestor. Study of the northern Rio Grande region suggests that the transition to centralized society and the shift from leaders toward rulers was underway and following different pathways during the fifteenth century.

These adaptations resulted from a variety of strategies. Independent villages included large, strong settlements such as Pecos, which dominated their smaller neighbors. Some individual villages like Tsiping were situated in such a remote location that they remained independent by staying hidden. Other villages confronted change by forming alliances, clusters of closely interacting villages that cooperated for economic, political, and social benefits. The focus of each alliance may have differed from place to place. In the Galisteo Basin protection of localized resources and trade routes would have been a priority, while in the Chama/Ojo Caliente subregion, local defense to protect land may have taken primacy, in an alliance outwardly indicated by weak bonds of trade (Ford 1972).

Experimentation in the nature of leadership was curtailed by the arrival of European explorers in 1540, and ended in the aftermath of the Pueblo Revolt, with the reconquest of the region by De Vargas in 1692.

Chapter **4**

Assessing Political Development in Copper and Bronze Age Southeast Spain

ANTONIO GILMAN

INTRODUCTION

Southeast Spain (Figure 4.1) has yielded a rich and distinctive later prehistoric cultural succession. Its essential features were defined over a century ago by the brothers Henri and Louis Siret, Belgian mining engineers in charge of the operations in the Sierra Almagrera in the province Almería. Their classic work, *Les Premiers Ages du Metal dans le Sud-est Espagnol* (Siret and Siret, 1887), and Louis Siret's continuing work in the region until his death in 1934, provided the basic corpus of information and the overall intellectual approach that informed the writing of the later prehistory of Spain until the 1970s. Within this framework the increasing social complexity exhibited by the sequence in southeast Spain was the product of diffusion from the eastern Mediterranean. In the past 25 years, this sequence has been reinterpreted from a variety of functionalist perspectives as a process of autochthonous evolution. The debates concerning the causes of this de-

---

ANTONIO GILMAN • Department of Anthropology, California State University, Northridge, CA 91330

From Leaders to Rulers, edited by Jonathan Haas. Kluwer Academic/Plenum Publishers, New York, 2001.

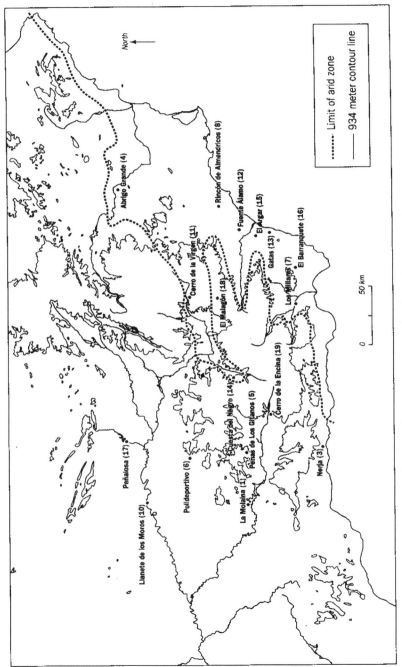

Figure 4.1. Neolithic to Bronze Age archaeological sites in southeast Spain.

ASSESSING POLITICAL DEVELOPMENT

velopment illustrate the challenges involved in assessing prehistoric economic and political institutions.

ENVIRONMENTAL BACKGROUND

Southeast Spain includes the most arid region of Europe. The coastal areas of the provinces of Almería and Murcia and some interior basins (such as those around Guadix and Baza in the province of Granada are in the rain shadow of the Betic mountain systems. Below 500 m elevation this area receives from 200 to 400 mm mean annual rainfall; in 9 to 11 months of the year there is a deficit between actual and potential evapotranspiration. In this arid zone, irrigation is a prerequisite for stable, productive agriculture; dry farming is only practiced opportunistically in the occasional wet year. As one moves west and north out of the rain shadow, precipitation both increases and becomes less variable, ranging from 400 to 600 mm and up to 1500 mm annually in the high mountains. In the moist zone, irrigation is a supplement to viable dry farming. These differences in climate within southeast Spain are stable and long term ones arising from the orographic pattern. Only changes in atmospheric circulation patterns greater than any that can plausibly be postulated for the Holocene would appreciably reduce the contrast between the arid coastal lowlands and the moist inland uplands. Accordingly, it seems clear that the later prehistoric inhabitants of southeast Spain would have confronted highly differentiated constraints with regard to their productive endeavors.[1]

ARCHAEOLOGICAL SEQUENCE

The main features of the later prehistoric sequence in southeast Spain may be briefly summarized as follows.[2]

The earlier Neolithic of southeast Spain is a variant of the Impressed Ware complex found throughout the western Mediterranean. Here as elsewhere, Impressed Ware sites are those of the first farming communities of the region. Almost all of the known sites are occupations in caves or rock shelters (hence its traditional name of *Cultura de las Cuevas* [Bosch Gimpera 1932]), but recent work has uncovered some open-air sites attributable to

---

[1] Vilá Valentí (1961) and Geiger (1972) provide concise summaries of the characteristics and economic implications of the present-day climate of southeast Spain (cf. Chapman 1978,1990: 98–105; Gilman and Thornes 1985:9–16; Stika 1998).

[2] Robert Chapman (1990:54–96) provides a comprehensive review of the evidence underlying this summary.

this period.³ Earlier Neolithic sites with reliable radiocarbon dates fall between 5500 and 4500 calBC.⁴ The most comprehensive survey of the *Cultura de las Cuevas* (Navarrete Enciso 1976) indicates that almost all known sites of this period are found in the moist uplands of southeast Spain, with only a few documented cases in the arid sector.

The interval between the Impressed Ware Neolithic and the Millaran Copper Age (4500 to 3500 calBC) is poorly known (and may be referred to as "later Neolithic" almost by default). In the moister sectors, there is occasional evidence for a continuation of occupation at some cave and rock shelter sites (with ceramic assemblages exhibiting reduced percentages of incised and impressed sherds), but the general dearth of evidence is perhaps best explained by a greater use of open-air sites, whose short-term occupations would have left little of archaeological salience.⁵ In the arid part of southeast Spain, the absence of any recently excavated, radiocarbon-dated sites belonging to this period makes matters even more confused. In their classic 1943 survey of the megaliths of southern Iberia, Georg and Vera Leisner assigned some of the simpler monuments with less elaborate grave goods to a period prior to the Millaran, and some settlements lacking traces of metallurgy or clear Copper Age-type fossils have also been assigned, speculatively, to a Later Neolithic (this would be the "Almería Culture" of Bosch Gimpera [1932]). The existence of this cultural facies has not yet been confirmed by more modern, radiocarbon-dated discoveries, however, and its contents may simply group together some of the less elaborated settlements and burial monuments of the following period. The arid sector of southeast Spain is one of the most intensively investigated regions of the Iberian Peninsula, and the failure, after well over a century of research, to find substantial traces of either an earlier or a later Neolithic occupation in the area suggests that the lacuna represents a reality of sparse, ephemeral settlement.⁶

---

³For example, La Molaina (Pinos Puente, Granada province) (Sáez and Martínez 1981) or Cerro Virtud (Cuevas del Almanzora, Almería province) (Montero Ruiz and Ruiz Taboada 1996; Ruiz Taboada and Montero Ruiz 1999).

⁴A number of sites with Impressed Ware Neolithic materials have produced radiocarbon dates in the later seventh and earlier sixth millennia calBC—in southeastern Spain, Cueva de Nerja (Nerja, Málaga Province) (Pellicer and Acosta 1986) and Abrigo Grande (Cieza, Murcia Province) (Cuenca Payá and Walker 1986)—but these present a variety of contextual problems (Zilhão 1993).

⁵The best-documented example of the continuation of the *Cultura de las Cuevas* is the site of Peñas de los Gitanos (Montefrío, Granada province) (Arribas and Molina 1979). A recent and well-documented example of an open-air establishment is the Polideportivo site (Martos, Jaén province) (Lizcano et al. 1992), consisting of a variety of subsurface features (pits, ditches) preserved below the plow zone. This site type is assigned to a *Cultura de los Silos*.

⁶Fernández-Miranda et al. (1993) review the extant evidence from the Vera basin, the base of the Sirets' operations. Cerro Virtud (see footnote 3) is an additional example of the problematical nature of that evidence: It has radiocarbon dates in the early fifth millennium calBC, but also a pottery fragment with copper slag adhered to its inner face.

ASSESSING POLITICAL DEVELOPMENT

The Copper Age "Los Millares Culture" has left abundant and well-characterized remains in both the moist and the arid sectors of southeast Spain. This phase of the sequence falls between 3500 and 2250 calBC.[7] Large numbers of long-term, open-air, often fortified settlements are known, whose preservation is the result of their stone-built houses and settlement walls. In terms of artifact technology, the principal new feature of this period is the widespread appearance of a simple copper metallurgy. The settlements are generally small villages of less than 1 ha. The type site for the culture, Los Millares itself (Santa Fe de Mondújar, Almería province), is in the arid sector. It may have an areal extent of as much as 5 ha (which would make it perhaps the largest settlement in Mediterranean Europe during the third millennium). With its three lines of fortifications, large nearby megalithic cemetery, and subsidiary occupations on the surrounding hills, Los Millares gives the impression of being a "central place" of some sort (although a single exceptional site cannot be adequate evidence for a settlement hierarchy).[8] The prevailing Copper Age burial rite consists of collective inhumations in natural or artificial caves or in megalithic chambered tombs. The burial places are near settlements and are sometimes grouped in cemeteries, the largest of which (at Los Millares) has some 80 megaliths. Grave goods consist of ritual fetishes and utilitarian items, some of which are rendered in valuable raw materials (e.g., ivory, imported from North Africa, shaped into sandals or copper cast as knives). There are significant disparities in the elaboration of tombs and in the wealth of grave goods between individual tombs within cemeteries (Chapman 1981), as well as between cemeteries as wholes. Such wealth differentials appear to be more marked in the arid than in the moist sectors of the southeast. Apart from the unique site of Los Millares itself, settlements in the arid and moist sectors of southeast Spain are not clearly differentiated in terms of their size or elaboration (Hernando Gonzalo 1987). It is, rather, in the funerary monuments and their grave goods that the arid sector's relatively greater social complexity is apparent.

The Bronze Age "El Argar Culture" dates from 2250 to 1500 calBC. It exhibits clear changes in settlement and burial patterns from the preceding

---

[7]The chronological boundary between the later Neolithic, itself poorly defined, and the Copper Age is (to say the least) fuzzy, but that between the Copper Age and the subsequent Argaric Bronze Age is quite sharply defined and suggests that the transition was an abrupt one (Fernández-Posse et al. 1996).

[8]Excavations at the Los Millares site complex were conducted in 1891 by Louis Siret and his assistant, Pedro Flores (Siret 1893), from 1953 to 1957 by Martín Almagro Basch and Antonio Arribas (Almagro and Arribas 1966), and from 1978 to 1983 by a team from the University of Granada (Arribas and Molina 1982, Arribas et al. 1983). These investigations have established the monumental character of the site, but not its changing configuration over the millennium over which it developed. As a result, it is difficult to say how large the site was in each of its various phases. Part of the problem, of course, is that the most recent work has only been published in preliminary reports.

Los Millares Culture. Settlements are generally new foundations, acropolis sites in extreme, defensive positions.[9] These sites are generally very small, constrained by the hilltops on which they typically are placed. Burials are no longer collective; rather they are individual interments under the floors of houses. Grave goods consist of the personal finery and possessions of the dead (weapons, ornaments, and so on) and exhibit clear wealth differentials. As in the preceding period, these differences appear to be more marked in the arid portions of the region. Compared to the Copper Age, Bronze Age metallurgical technology is somewhat more elaborate (occasional tin bronze alloys, silver production, and two-part molds are Argaric novelties).

TRADITIONAL INTERPRETATION

The principal features of this sequence were recognized by the Sirets over a century ago, and for almost a century after their initial work prehistorians agreed on how to explain its successive changes.[10] Working within a normative framework (whose essential features were also implicit in the Sirets' work), a framework in which different kinds of archaeological residue are seen primarily as the result of the different ideas that guided their manufacture, prehistorians interpreted changes in archaeological cultures as the result of changes of ideas, mostly brought about by influences from abroad. The sources of the new ideas were, of course, the more advanced centers of civilization in the eastern Mediterranean. Oriental beliefs and technologies (transmitted by agents described, depending on the predilection of the prehistorian, as colonists, missionaries, prospectors, or traders, or, more cagily, "influences") would have transformed prehistoric Europe, much as European civilization had transformed the world under capitalism. Virtually every significant change, from the introduction of farming and pottery-making in the early Neolithic to the development of megalithic burial rituals and metallurgy in later times, was explained by the direct intervention of external agencies. Scholars might disagree on the routes of diffusion (from Egypt by way of North Africa or from the Aegean by way of Italy) or on its agents (settlers, prospectors, and so on), but not on the overall thrust of the explanatory argument. The prehistory of southeast Spain was universally and straightforwardly interpreted as a clear example of what Childe

---

[9]There are also sites in low-lying areas, such as Rincón de Almendricos (Lorca, Murcia province) (Ayala Juan 1991:49–176).
[10]Martínez Navarrete (1989) provides a comprehensive review of the history of archaeological thought concerning the later prehistory of Spain.

(1958b:70) termed "the irradiation of European barbarism by Oriental civilization."

PROCESSUAL ALTERNATIVE

The *ex Oriente* lux explanation of cultural developments in Iberia had always had certain weaknesses. As Childe himself recognized,[11] the similarities between East and West involved broad typological parallels, not specific resemblances: Not a single undoubted eastern Mediterranean manufacture has been found in the Iberian Peninsula prior to 1500 BC (in spite of the fact that such finds were predicted and sought after during a century of research).[12] Furthermore, the general trait complexes found together in Spain and held to be derived from the east do not form assemblages in their supposed homelands. Such was the overall attractiveness and coherence of the Orientalist approach, however, that these difficulties were passed over until radiocarbon dating began to be applied to the Iberian sequence. In his classic paper, "Colonialism and Megalithismus," Colin Renfrew (1967) made it clear that allegedly Oriental features in Copper Age Spain were found in contexts that were as early as, if not earlier than, those of their prototypes in the Aegean and elsewhere. Renfrew, however, did not propose an alternative scenario that made sense of the Iberian sequence's development. Expectably, therefore, most prehistorians of Iberia found it possible to ignore the mere facts that Renfrew had pointed out.[13] Facts don't kick: What was needed was an alternative account of the sequence that could replace the normativist orthodoxy.

---

[11]"If prospectors and merchants from the Aegean helped to found Bronze Age colonies, they did not bring a complete material and ideological equipment with them, or maintain contacts with the homeland to keep them supplied with its manufactures as did the historical Greek colonists" (Childe 1958a:117–118).

[12]The temptation to find parallels is virtually impossible to resist. As prudent a scholar as Jean Guilaine (1994:290–291), for example, has cautiously revived classic diffusionist speculations (Almagro 1962) about the possible Near Eastern derivation of the javelin points found in the nineteenth century excavations of the Cueva de la Pastora passage grave (Valentina del Alcor, Sevilla Province). Montero Ruiz and Teneishvili (1996) have demonstrated, however, that the typological matches with Near Eastern specimens are not precise ones and that the metallurgical properties of the Pastora specimens match local Copper Age production practices exactly. When, at long last, a bona fide Aegean Bronze Age import *was* discovered in the Iberian Peninsula—the Late Helladic IIIA/IIIB sherd from the late Bronze Age site of Llanete de los Moros (Montoro, Córdoba) (Martín de la Cruz 1988)—it completely post-dated the entire Millaran–Argaric cycle.

[13]As one archaeologist put it in the early 1970s, "a mere few C-14 dates from at most five sites do not constitute adequate grounds on which to presume to change the origin of the entire megalithic culture of the west" (Almagro Gorbea 1973:198).

The most parsimonious alternative to the diffusionist narrative of prehistoric culture change in southeast Spain would be an evolutionist one, and Robert Chapman and I, the first processually oriented archaeologists to review the case of southeast Spain systematically, independently developed coincident accounts of its later prehistoric development. (These accounts had different theoretical twists, as will be discussed below, but they involved substantially similar readings of the available evidence.)

The Neolithic–Copper–Bronze Age sequence presented clear indications of social change. The main features of the cultural succession could be straightforwardly interpreted as the result of (1) increasing craft specialization (seen primarily in the development of metallurgy), (2) increasing divergences in wealth within and between communities (seen in the changing character of mortuary ritual and the increasing differentiation of grave lots), and (3) increasing militarism, both as a practice (seen in the development of settlement fortifications) and as a value (seen in the weaponry of Bronze Age burials). If these tendencies, which mainly involve the social organizational aspects of the prehistoric record, could be linked to changes in production systems over the course of the sequence, one would be able to explain the salient transformations of the southeast's prehistory as a product of local, autonomous evolutionary development, without recourse to dubious external agencies.

Now, it is important to realize that in the early 1970's, when Chapman and I began to develop our interpretations of the later prehistoric sequence in southeast Spain, evidence for changes in the economic base of prehistoric change in Iberia (evidence required for an evolutionary explanation) was largely unavailable. The situation represented, as it were, a reversal of Christopher Hawkes' (1954) famous inferential hierarchy: we knew far more about the superstructure than about the infrastructure of the Millaran and the Argaric. In the older archaeological literature on southeast Spain, the economic aspects of the prehistoric sequence were given little attention. Early excavations had shown that by the Neolithic the inhabitants of the southeast were farmers; further study of subsistence practices and man–land relations were thought unnecessary, since it would only reveal mundane common denominators of little relevance to the human prehistory that mattered. Development of an evolutionary approach would require research into the economic underpinnings of the social changes apparent in the prehistoric record.

The starting point for this research was a realization of the significance of southeast Spain's ecological contrasts, contrasts that went almost unmentioned in the literature. It seemed clear that in the past, as now, subsistence farmers in the interior and in the coastal zones of southeast Spain would have operated under very different environmental constraints. At the same

ASSESSING POLITICAL DEVELOPMENT 67

time, it seemed clear that the tendencies towards social complexity that were enumerated above were stronger in the arid coastal lowlands of southeast Spain than in the moist uplands of the interior. Connecting these two facts seemed like an obvious approach to the construction of an evolutionist account of the region's later prehistoric sequence. Archaeology, after all, is a discipline in which it is good to be obvious.

From the Copper Age on, much of the evidence for the prehistoric sequence in the arid southeast comes from long-term fortified settlements, whose inhabitants must have practiced a stable agriculture. Chapman and I were led to conclude that, in a region as marginal for dry farming as the arid zone of coastal southeast Spain, irrigation must have been a critical feature of the prehistoric production system. Excavated evidence for this proposition was not entirely lacking. Thus, the Sirets had recovered fragments of textiles made of flax from Argaric sites in the arid zone, suggesting cultivation of a crop whose demand for water would make it most unlikely to have been dry-farmed under conditions anything like those of the present (cf. J. M. Renfrew 1973:34). In the 1960s Wilhelm Schüle's excavations at the Copper Age site of Cerro de la Virgen (Orce, Granada province) had even revealed a possible irrigation ditch (Schüle 1967), which, of course, he interpreted as another example of Oriental influences in Iberia. Irrigation is a form of agricultural intensification often implicated in the development of social complexity elsewhere in the world, and it could be placed at the base of such a development in southeast Spain as well.

The evidence was sparse, however, particularly when one considered the theoretical weight Chapman and I proposed to place on it. To strengthen that argument, Chapman in his 1975 doctoral dissertation and in the publications which emerged from it (Chapman 1978, 1981, 1982, 1990) undertook a detailed functionalist examination of the available evidence and showed how it supported both the irrigation hypothesis and the overall pattern of emergent complexity over the course of the sequence. After writing some preliminary, programmatic articles (Gilman 1976, 1981), I undertook a program of site catchment analysis for all the better-documented published Neolithic to Bronze Age sites in both arid and humid zones of the southeast (Gilman and Thornes 1985), so as to provide a geographical test of the irrigation hypothesis.

It is our duty as would-be scientists to attempt to develop evidence for our arguments, and it is satisfactory when that evidence tends to confirm them, but only the most innocent would suppose that the evidence carries the day for the arguments. Before any substantial new evidence (such as the site catchment analysis) on behalf of an evolutionist reading of the sequence had been developed, our programmatic manifestos had elicited functionalist counterarguments from Spanish colleagues (e.g., Ramos Millán 1981).

The rapidity of the change of opinion is exemplified by the presentation to the 1984 Siret memorial conference of a positive, attributed summary of Chapman's and my arguments by Antonio Arribas Palau (1986), the excavator of Los Millares and a long-term proponent of its colonial origin.[14] Although some scholars would continue to support the diffusionist paradigm (e.g., Schüle 1986; Eiroa 1989; González Prats et al. 1998), the debate between functionalists and normativists ended before it began and was replaced by a debate among functionalists about which of their explanatory sketches had most merit. Discussion has revolved around two separate issues. One has to do with the relative weight of agricultural intensification or metallurgy in the development of social complexity in the southeast. The other concerns the character of that complexity, that is to say, the extent to which the concentration of wealth in limited social segments was the result of exploitation.

METALLURGY

An approach of long standing in discussions of the origins of social inequalities emphasizes the role of metallurgy in stimulating commodity exchange (e.g., Childe 1954: 87–88; cf. Engels 1972: 233). Vicente Lull (1983) adapted this theory to the Argaric Bronze Age in order to construct a nondiffusionist theory of its development:

> The development of metallurgy . . . produced a sharp change in . . . production which in turn brought about and made necessary other changes in social relations. The original self supporting communities became communities with complementary production requiring trade. This also caused an improvement in communication and routes of transport; an improvement that fostered a managerial hierarchy, most likely for security reasons. (Lull 1984: 1222, 1983: 456)

To the extent that miners and smiths are full-time specialists, their production must exchanged with farmers and other producers to meet the specialists' needs. To the extent that farmers and others depend on metal tools for effective production, they must engage in exchange with metalworkers. Thus, the mutual dependency of specialist producers leads to the development of internal and external trade networks, which in turn provide opportunities for middle-men to provide entrepreneurial and managerial services and/or to establish economic control. This theory proposes, then, that (in Haas's [1982] terms) the cost to producers of complying with the middle-

---

[14]For a discussion of the circumstances that brought about the ready acceptance of functionalist perspectives in Spain in the late 1970s and early 1980s, see Gilman (1995b).

ASSESSING POLITICAL DEVELOPMENT

men's retention of a surplus would be less than the costs of refusing to do so (namely, deprivation of indispensable supplies).

This theory is an archaeologically testable one. If it is true, one might reasonably expect (a) that metal production would be an activity carried out by skilled specialists in workshops distinguishable from ordinary domestic production areas, (b) that the metal production would meet basic needs by providing better cutting implements deployed in primary (e.g., agricultural) production activities, and (c) that the scale of production and exchange of metal goods would be sufficiently large to make the recipients of these necessities dependent on their suppliers. The general character of the southeast's archaeological record had generated substantial skepticism about such claims in some quarters (e.g., Chapman 1984, Gilman 1987a), and the recent detailed survey of Copper and Bronze Age metallurgy in the provinces of Granada, Almería, and Murcia conducted by Ignacio Montero Ruiz (1994, 1999) indicates that such doubts were justified. Montero shows that:

a. Smelting of ores was carried out at inefficiently low temperatures in small batches in ordinary-seeming domestic contexts.[15] Metalwork shows little evidence of highly developed skills: molds are extremely simple and alloys are rarely deliberately controlled. This seems to be consonant with the part-time, kin-based, dispersed end of Costin's (1991) scale of production organization.

b. There are no agricultural implements known from either the Copper or the Bronze Age.[16] Copper artifacts interpretable as tools consist mainly of awls. In the Argaric over three quarters of the metal objects catalogued by Montero are weapons and ornaments. This suggests that the primary function of the industry was nonpractical. In Haas's (1982) terms, the material costs of refusal to participate in the metallurgical system would have been minimal.

c. The overall scale of production is tiny. The total number of metals recovered from Montero's study area is under 600 and 3000 for the Copper and Bronze Ages, respectively (Montero Ruiz 1994: 213).[17] Furthermore, the trace-element signatures of slags and finished arti-

---

[15] Reference has been made to a house at Los Millares exclusively devoted to metallurgical production (Molina González 1988:261), but the contextualized evidence that might support such claims remains unpublished.

[16] The 59 Copper Age and 139 Bronze Age copper axes known from the three provinces (Montero Ruiz 1994:213) would not made much of an impact on land clearance over the course of two millennia, even assuming they were used for that purpose.

[17] The quantitative poverty of the Argaric Bronze Age is particularly striking in comparison to other regions of Europe. Montero lists a total of ten swords for the three province he surveys. In Denmark, where all bronze is imported by long-distance trade, the same number of Bronze Age swords can be found in a few parishes (e.g., Aner and Kersten 1984:3–8).

facts are different from site to site. This suggests that exchange and recasting of metal was minimal.[18] It also suggests that each village obtained its metal from local sources of varying composition, a pattern like that documented by Barrera Morate et al. (1987) for stone axe procurement in the southeast.

Until further evidence is developed, it would appear that the production, exchange, and use of copper and bronze in southeast Spain did not create the dependencies postulated by the commodity-exchange explanation of emergent social stratification.[19] Metallurgy may have served to concentrate and display wealth, but apparently it was not the origin of that wealth.

AGRICULTURAL INTENSIFICATION

As mentioned earlier, the arguments Chapman and I had put forward made agricultural intensification (and, in particular, the development of irrigation systems)—not metallurgy—a key factor in the social evolution of later prehistoric southeast Spain. Now, direct evidence for irrigation in the prehistoric southeast is necessarily sparse. Irrigation is typically an off-site activity, the traces of which will not be found in archaeological excavations, and in the countryside more recent hydraulic facilities will have obliterated older ones. The results of the site-catchment analysis mentioned were consistent with the proposition of irrigation was important in the Copper and Bronze Ages, but that site-catchment approach is necessarily a circumstantial one, which a skeptic may doubt. Likewise, paleobotanical finds showing the presence in ancient sites of species that now must be irrigated can be explained away by the conjecture that such water-demanding species were cultivated because the climate of the Copper and Bronze Ages in the southeast was moister than it is at present.

This, in fact, is the approach taken by most scholars who reject the hypothesis that the development of irrigation was an important factor in the southeast's prehistoric development. They are constrained to argue that the

---

[18]Lull and Risch (1996:106) state that the lead isotope analyses conducted by Stos-Gale and others (and reported in an as yet unpublished communication to the Junta de Andalucía) indicate that metal artifacts from Argaric sites of Fuente Álamo (Cuevas del Almanzora, Almería province) and Gatas (Turre, Almería province) derive from sources near Jaén, but Montero Ruiz (1999:351) indicates that this lead isotope evidence shows nothing of the kind.

[19]In his more recent publications Lull at first neither cited nor confronted Montero's arguments (e.g., Lull and Risch 1996), and subsequently he has stopped using the metallurgical exchange model to explain Argaric development (e.g., Castro et al. 1998).

ASSESSING POLITICAL DEVELOPMENT

region's climate during the Copper and Bronze Ages was significantly more humid than that of the present (so that such an intensification would be unnecessary for the stable agricultural production that the prehistoric inhabitants of now-arid areas clearly practiced).

Palynological evidence from the southeast is not abundant (Florschütz et al. 1971, Pons and Reille 1988; Burjachs and Riera 1996, Pantaleón-Cano et al. 1996), but it suggests that as of 6000 years ago, the main changes in the vegetation pattern are the result of human intervention.[20] Accordingly, proponents of a more humid climate in the Copper and Bronze Ages rely primarily on the faunal and macrobotanical evidence from archaeological sites. Much has been made (e.g., by Molina González 1983:71–72), for example, of the presence of species such as otter at Cerro de la Virgen (von den Driesch 1972:125) and beaver at Cuesta del Negro (Purullena, Granada province) (Lauk 1976:84), even though these sites today have year-around streams flowing in their immediate vicinity. At Los Millares, the presence of water turtles among the fauna (Peters and von den Driesch 1990:64) and of fragments of alder, ash, tamarisk, and willow among the charcoal (Vernet 1997:153) suggests that the river Ándarax (which the site overlooks) would have flowed year around, but this is not inconsistent with a climate similar to that of the present: The Ándarax drains the entire southern flank of the Sierra Nevada and would have a year-around flow today if its water were not completely abstracted for irrigation.

The available palynological and anthracological evidence (Rodríguez-Ariza 1997; Vernet 1997) is consistent, furthermore, with reconstructions (Freitag 1971) based on relict stands of undegraded vegetation, of what the region's plant communities would be like in the absence of human denudation. The critical point is that the undoubtedly lesser degree of desertification in the prehistoric past (the presence of chaparral or forest on now bare hillslopes, greater ground-water infiltration, springs with more abundant flows, dry washes with more regular stream regimens, presence of now locally extinct game species, and so on) would not affect the possibilities for dry farming (cf. Stika 1998). Such farming depends on rainfall, and the region's topography would be bound to create significant differences in this regard. Climatic oscillations during later prehistoric times would have been insufficient either to erase the contrasts between more humid and more arid sectors of the Peninsula or to diminish significantly the advantages which an intensified agriculture (including irrigation) would present in the less-favored sectors.[21]

---

[20]This is not to say that there were no climatic fluctuations over these millennia. Interestingly, the most recent analyses suggest an increase in aridity at about 4500 years ago (Pantaleón-Cano et al. 1996:32), in other words at about the Millaran–Argaric transition.

[21]It is worth noting that in 1953 W. L. Kubiena indicated that the xerorendsina soils at Los

Recent carbon discrimination analysis of cereals and legumes from a number of recently excavated Copper and Bronze Age sites in arid sectors of southeast Spain (Araus et al. 1997) provides a more direct test of the irrigation hypothesis. Cultivars that receive sufficient water during grain filling show greater discrimination against ^{13}C than those that suffer decreased water availability and increased evapotranspiration. Wheat and barley from these localities (Los Millares and Cerro de la Virgen are among them) exhibit discrimination values below what one would expect for well-irrigated crops, while fava beans show somewhat higher values. This suggests that, if irrigation was practiced, it was limited in scope or inefficient in its timing (as might be the case for floodwater farming).

Paleobotanical results from recent excavations are also consonant with a limited hydraulic agriculture. At the Argaric sites of Gatas, under excavation since 1986 by Chapman, Lull, and their associate; at Fuente Álamo, under excavation since 1977 by the Deutsches Archäologisches Institut; and at the type site of El Argar itself (Antas, Almería province), tested by the Deutsches Archäologisches Institut in 1991, by far the dominant cultivar is barley (Stika 1988; Stika and Jurich 1998; Castro et al. 1999). Of the crops documented from later prehistoric southeast, barley is the most suitable for dry farming. Flax and fava beans are found at all of these Vera-basin sites, however, and these are crops that require more waater than the rainfall available in the region would provide. Small artificially watered garden plots must therefore have existed in Bronze Age times.

The agricultural improvements that the inhabitants of southeast Spain invested in over the course of the Copper and Bronze Ages go beyond possible irrigation practices, however. Thanks primarily to the work of Joachim Boessneck, Angela von den Driesch and their collaborators, we have substantial numbers of faunal analyses from the southeast. These suggest a pattern of progressively intensified exploitation of animals over the course of the sequence.[22] Neolithic faunal series emphasize sheep, goat, and pig (species apparently exploited for their meat), while in the Copper Age, the proportion of cattle increases greatly. In the later Copper Age and the Bronze Age, some sites have considerable proportions of horse. Horse and cattle are species useful for their traction as well as their meat, but their breeding and their maintenance as adults requires a substantial investment of human work for a partially deferred return. If, as elsewhere in later prehistoric Europe, this shift in animal husbandry is accompanied by the

---

Millares were formed under climatic conditions similar to those of the present (Almagro and Arribas 1963:261), a condition confirmed by isotope analysis of the calcareous crust on which the settlement was built (Capel et al. 1998).

[22]See Chapman (1990:115–118) for references.

introduction of the plow, even more extensive investments in the land may be inferred. That the faunal assemblages from Copper Age collective tombs[23] have even higher proportions of cattle than the assemblages from contemporaneous settlements suggests that these animals were invested with a ritual significance appropriate to their long-term value as instruments of production. There is substantial evidence, then, that Sherratt's (1981) "secondary products revolution" took place over the course of the Millaran and the Argaric.

Comparatively reliable paleobotanical studies of plant remains based on flotation of sediments are only beginning to be published from the southeast, but the older available evidence suggests a greater variety of cultigens in the Copper and Bronze Ages than in the Neolithic (Chapman 1990:114–115). More recent studies contain some intriguing hints of intensified cultivation practices as well. There is some evidence for incipient tending of olives in eastern Spain as early as the Impressed Ware Neolithic (Barton et al. 1990; Terral 1996). Anthracological evidence from Los Millares provides some evidence for olive cultivation. Olive is the most frequent taxon recovered among the charcoal fragments (Rodríguez-Ariza and Esquivel 1990). Furthermore, analysis of the growth rings in olive fragments indicate the exploitation of trees that grew relatively rapidly, a condition that today would be attributed to domestication (Rodríguez-Ariza and Vernet 1992:5). Likewise, at Gatas the concentration of olive charcoal is greater than what would be predicted from reconstructions of the natural vegetation in the site's vicinity (Ruiz et al. 1992:22). Palynological evidence makes it clear, however, that extensive plantations of olives did not come into existence until history times (Pantaleón-Cano et al. 1999:21).

Although much more needs to be done, particularly in the area of paleoethnobotany, the available evidence indicates that agriculture became somewhat more intensive over the course of the later prehistoric sequence.

MANAGEMENT OR EXPLOITATION?

The evidence developed for certain intensification of subsistence production over the course of Copper and Bronze Ages of southeast Spain confirms the relevance of an evolutionary approach to Millaran–Argaric development. There is, however, considerable disagreement about how the changes in agricultural production would articulate with the major social transformations (the intensification of warfare and the increase in wealth

---

[23]The fauna from the megalithic cemetery of El Barranquete (Níjar, Almería province) (Almagro Gorbea 1973) was studied by von den Driesch (1973).

disparities) over the course of the sequence. An association between agricultural intensification and social inequality occurs in all evolutionary sequences towards greater social complexity. Leaders ambitious for themselves and for their followers—Clark and Blake's (1994) aggrandizers—exist in all societies. In societies with extensive systems of production their ambitions are usually frustrated, but in societies with intensified agricultural production leaders succeed in establishing hereditary control: They become rulers and their followers, subjects. As Haas (1982) has made clear, there are two basic, contrasting explanations for this correlation.

The first approach—and until recently the dominant one among anthropological archaeologists—underlines the functional integration required by intensified production and the higher populations to which such productions give rise. A kin-based social organization would be unable to undertake the management of the requisite investments and specializations. In Marshall Sahlins's (1972:140) words, "the chief creates a collective good beyond the conception and capacity of the society's domestic groups taken separately. He institutes a public economy greater than the sum of its household parts."

The second approach, which in recent years has been gaining ground in our profession (e.g., Price and Feinman 1995), stresses the internal conflicts characteristic of complex societies. The question becomes "How do bosses get away with it?," not "What good are bosses?" In this view, the exploitation which underlies permanent inequalities within a society is made possible because the development of intensive production systems permits the reliable collection of tribute.

Given the empirical weakness of claims for large-scale management-requiring or dependency-inducing metallurgical production, the most promising economic cause of the social inequalities in southeast Spain is agriculture. Here differing scenarios for developments in southeast Spain have been put forward by proponents of integration and conflict theories of the emergence of stratification. Clay Mathers (1984) and, more cautiously, Chapman (1990:211–219) have suggested that leaders acted as redistributors, organizers of regional exchange systems in foodstuffs (for the facilitation of which metal and other valuables would serve as a form of primitive money), and regulators of access to restricted resources (such as water for irrigation systems). The managerial explanation of the emergence of inequalities assumes that the scale of the economic system is so large that permanent leadership is required for its successful operation. On this account, then, the individuals found in the wealthier burials of the Copper and Bronze Ages would have been administrators whose services helped the general population to stabilize the uncertainties of production in the high-risk envi-

ASSESSING POLITICAL DEVELOPMENT

ronment of southeast Spain. They would have organized the stockpiling and exchange of food to ward off local agricultural failure, adjudicated disputes over water and other scarce resources, and so on. Thus, the cost to primary producers of participating in the public economy (payment of a part of their product to a leader) would be compensated by the benefits such a participation would bring them.

Criticism of the applicability of the managerial account of social stratification to European prehistoric societies has centered on two questions: (1) Would the scale of the economies documented in Copper and Bronze Age times require permanent managers? and (2) Is the function of incipient aristocracies managerial, in any event? I have argued elsewhere (Gilman 1981, 1987b, 1991, 1995a) that analogous historical and ethnographic cases indicate that the answer to both questions is no, and I do not propose to repeat those arguments here. Instead, I would like to discuss the main evidence deployed to support the managerial account, namely, the existence of site hierarchies. Following Gregory Johnson (1973), integrationists have considered multiple-tiered site hierarchy to be a critical diagnostic of social stratification: A center of economic and political organization should be larger and more monumental than villages of primary producers, for the resident chief will attract subordinates freed from direct productive activities to do his bidding. This idea was imported to prehistoric Europe by Renfrew (e.g., 1973b) and Chapman applies the idea to the Millaran and Argaric:

> [Settlement] size data . . . support the inference of an increase in political centralisation in the later third and second millennia B.C. in south-east Spain, with the emergence of a two-level hierarchy. Such a hierarchy has been thought, by supporters of social typologies, to support the inference of a chiefdom society. (1990:176)

Unpublished systematic site surveys by the University of Granada suggest to Molina González (1998:259) that "The numerous Copper Age sites are distributed in an unstructured fashion," but that "during the Bronze Age there would arise a regular site distribution with a clear hierarchy and functional differentiation."[24] Whether such claims are applied to both the Copper and Bronze Ages or to the Bronze Age only, the critical problem they present is the restricted range of site sizes: The largest sites (around 5 or 6 ha, not necessarily completely occupied at any given time in their history) scarcely attain the level of towns. When villages (covering a few hectares) exist side by side with farmsteads (covering a fraction of a hectare), the resultant

---

[24]Publications to date of these surveys (Maldonado Cabrera et al. 1992, Moreno Onorato et al. 1992) do not provide the site size data required to evaluate Molina's assertions, however.

settlement hierarchy need not reflect a social hierarchy.[25] All in all, the evidence tends to support Mathers' (1994:54) conclusion that "well defined settlement hierarchies" are absent.

The advantage of the conflict-oriented approach to stratification is that it does not suppose that the small-scale farming evidently characteristic of the Millaran and Argaric required higher-order regulation. The development of intensive systems of cultivation would have changed the social structure of southeast Spain not because they demanded management, but because the capital investments such systems involved opened up the possibility of effective exploitation of the cultivators (Gilman [1976, 1981] and Vicent García [1995] following Childe [1951:90] and Adams [1966:54]). The investments would increase long-term productivity of the tracts to which they had been applied, and farmers would be correspondingly reluctant to abandon them for unimproved land. Easy group fission is the essential mechanism by which an egalitarian social order is maintained as such. If members of a society do not depend on capital investments to produce what they need, they can abandon aggrandizers when their ambitions become intolerable.

After the primary producers in southeast Spain intensified their production beyond a certain point, however, they would no longer be able to do so: As Michael Mann (1986) would put it, they had become caged. Agricultural intensification would have provided leaders the leverage with which to become rulers. The "previous accumulation" (Vicent García 1995:178) of capital investments villagers had built up would require defense, but their very value would make it difficult to control those charged with their defense. The efforts of cultivators to improve their material security by making long-term improvements in the land would eventually undermine their social security by making them vulnerable to a protection racket. On this

[25]In any event, the size of larger sites has not been evaluated against the agricultural productivity of their catchments, as Brumfiel (1976) and Steponaitis (1978) recommend. It is interesting, for example, that Los Millares overlooks the Ándarax, the largest watercourse in the arid southeast. A more persuasive argument might be constructed on the basis of functional differentiations between sites: A central place should be the site of chiefly institutions and practices that would leave signatures in the archaeological record that would be absent in egalitarian villages. Thus, the lines of city walls that close off the Los Millares promontory seem more monumental than the fortifications of other Copper Age sites. Likewise, the freestanding thick-walled rectangular buildings, O and H, on the uppermost platform of the Argaric site of Fuente Álamo are not found at other excavated sites and might be interpreted as public edifices of some sort. The excavators' interpretation of what these buildings were used for is indecisive (Schubart et al. 1985:78), which suggests that clear evidence of their function is lacking. These possible examples of chiefly facilities are isolated ones and, given the uneven quality of the extant archaeological record, it would be unwise to make too much of them.

ASSESSING POLITICAL DEVELOPMENT

account, then, the Millaran–Argaric transition would have taken place when the costs of compliance to leaders' demands for surplus became less than the costs of refusal (relinquishing access to productive resources) and the leaders could make themselves rulers and their followers subjects.

RULERS OR LEADERS?

This whole scenario assumes, of course, that the social inequalities reflected by the substantial wealth differentials[26] in Bronze Age burials were hereditary in character. This assumption has been shared by all processualists who have worked on the sequence in southeast Spain. In terms of sociopolitical typologies, the Argaric would at least be a chiefdom, and in the view of some (Arteaga 1992; Lull and Risch 1996), even perhaps a state. This last position, if I understand it correctly, does not seem to involve a classic, governmental definition of the state. It is based, rather, on a dislike of the way chiefdoms have been interpreted by systems functionalists (Nocete 1994). That is to say, the existence of hereditary social stratification is considered a sufficient criterion for the existence of a state, with no necessity of formal military, fiscal, or religious institutions, none of which are clearly apparent in Iberia before the Iron Age (Vicent García 1995:181). Lull and Risch (1996) affirm, of course, that the stable exercise of power requires economic and ideological control, but they present no evidence that this control was institutionalized.[27] Thus, when Iberian Bronze Age "states" come to be compared with their contemporaries in, say, the Near East, their proponents will be constrained to distinguish between states of the sort found in barbarian Europe and states sensu stricto.

Be that as it may, even the more moderate view that the Argarics were at the chiefdom stage faces the difficulty that the evidence for hereditary stratification is hardly clear. The proposition that the wealthier Argarics were petty aristocrats is based in essence on a contextual interpretation of the

---

[26]Lull and Estévez (1986) argue that Argaric burials can be ranked by wealth into five strata.
[27]The existence of hilltop citadels (Fuente Álamo is the case in point) that are large in proportion to the agricultural resources in their vicinity and that have evidence for large scale grain milling (Lull and Risch 1996:104–105; Risch 1998) can plausibly be interpreted as evidence that their inhabitants collected a surplus from agriculturalists living in smaller, more conveniently located sites (although other readings of the evidence are possible). However, this in no way implies that this surplus was collected by tributary institutions as taxes. Likewise, the unquestionable stylistic uniformity of Argaric ceramics, metalwork, and burial practices suggests that the inhabitants of Argaric sites had mental templates in common, but not that these templates were imposed by ideological institutions as a form of thought control (Lull and Risch 1996:107).

patterns of change from the Millaran to the Argaric, in particular the change from collective burials with grave goods consisting of primarily functional or ritual items to individual burials with grave goods consisting of weapons and ornaments. This, associated with evident exacerbation of intercommunity violence, can be packaged plausibly with the evidence for agricultural intensification to produce the explanatory sketch outlined above. If the superordinate statuses reflected in rich burials were achieved and not ascribed, however, the concentration of wealth could be quite free of any implication that it was accumulated exploitatively. A number of lines of evidence cast doubt on the processualists' consensus.

A skeptic might well ask, for example, why there is no evidence for residential differentiation in the Argaric to correspond to its mortuary differentiation: Aristocrats typically feather their nests, not just their graves, and the Argaric practice of burying their dead under house floors should enable one to pair up rich grave goods and high living standards. That scant advantage has been taken of this opportunity does not suggest that such correspondences are evident. It is true, of course, that functionally-oriented excavations in southeast Spain have been undertaken only in the last 20 years, and that their publication remains confined to preliminary or partial reports, so that it is difficult to evaluate systematically differential patterns of consumption within and between sites.[28] All the same, prudence is not a char-

---

[28]Perhaps the best example of household archaeology at an Argaric site is provided by the excavations at Peñalosa (Baños de la Encina, Jaén province) (Contreras Cortés et al. 1995). The architectural remains do not seem to be strongly differentiated across the site's three terraces, and evidence of grain storage and metal production occurs in most structures. Furthermore, richer and poorer burials are found within the same structures. It is claimed that the distribution of faunal elements in the various habitation unites shows differences between "richer" units in the higher areas of the site and "poorer" ones in the lower areas (Contreras Cortés, Morales Muñis et al. 1995), but the frequency of the minimum number of indivudals of the main meat-bearing animals (horse, cattle, sheep/goat, pig, and deer) across the site does not exhibit statistically significant contrasts. On the whole, the work which excavators have published concering their work does not self-evidently support their conclusion that the site was occupied by households belonging to different social classes.

Ramos Millán's (1997) spatial analysis of lithic manufacture in the second (late Copper Age) phase of occupation at El Malagón (Cúllar-Baza, Granada province) illustrates the difficulty of drawing social conclusions from limited evidence. The main excavated area (Ramos Millán: Fig. 2) reveals six round huts, of which the central one, with about 20 m^2 of internal space, is about four times larger than the other five. The large hut has a statistically significant greater ratio of finished lithic artifacts to débitage than its smaller neighbors, and Ramos Millán interprets this, not unreasonably, as evidence that it was inhabited by a "big man" who organized lithic exchange between settlements. Given the limited number of finished artifacts (14 in all) found in the large hut, it is not clear to me, however, why this distribution might not be interpreted as the result of differentiated activities and deposition within communal living quarters (as in the contemporaneous "Fortín 1" at Los Millares [Molina González et al. 1986]) without any imputation of political/economic ranking.

acteristic archaeological virtue. When our colleagues think they see patterns in their evidence, they are only too happy to tell us: The scarcity of claims concerning wealth differentials in domestic contexts speaks to their absence. As DeMarrais et al. (1996) point out, for hereditary elites conspicuous, archaeologically salient consumption is a strategy of power.

Another potentially fruitful line of evidence for class differentiation would be biological differences in skeletons. A hereditary ruling class, if it existed, might be expected to be better fed as children and longer lived and healthier as adults than their subjects. Manfred Kunter's (1990) survey of the surviving skeletal remains from the Sirets' excavations documented relatively few cases of pathologies related to disease or dietary stress,[29] which does not support the view that differences in consumption were strongly marked. The study of Jiménez Brebeil and García Sánchez (1990) of the physical characteristics of 19 skeletons from the Argaric site of Cerro de la Encina (Monachil, Granada province) indicate that individuals with lesser muscular development and fewer instances of arthrosis and periostitis were found in tombs with richer grave goods or more "important architectural constructions" (Jiménez Brebeil and García Sánchez 1990:174), but they give not detailed description of the archaeological association, so that it is difficult to evaluate their suggestion that the evidence "seems to indicate a probable social differentiation" (Jiménez Brebeil and García Sánchez 1990:174). The only instance of cribra orbitalia comes from one of the larger tombs, however. Jane Buikstra has conducted the physical-anthropological study of the skeletons recovered in the recent excavations at Gatas. The results published so far (Buikstra et al. 1995:166–167) with respect to pathologies related to diet and disease stress are shown in Table 4.1: They are manifestly inconclusive. We await the reports on stable isotope analyses, frequency of Harris Lines, and so on, but the results so far show little patterning.

Another facet of the archaeological record from southeast Spain which should trouble those of us who have espoused the existence of hereditary social stratification during the Argaric Bronze Age is less direct than the absence of evident differentiation in living standards. I refer to the extreme scarcity of ritual objects or sacred structures. As we have noted (Martín et al. 1993:40; cf. Lull, 2000), very little is recovered from Argaric sites that cannot be interpreted in technological or political economic terms. Practical utility, social ostentation, and political competition can explain virtually all of the variability of artifactual and architectural assemblages, with little left that would require us to interpret the record ideologically.[30] Rousseau (1947:172)

---

[29] Of the 793 individuals sufficiently complete to be assigned an age at death, only 63 exhibited enamel hypoplasias and just 2 cribra orbitalia (Kunter 1990:88, 95).

[30] It is notable, for example, that the Argaric is hardly addressed in any of the numerous articles dealing with Iberian materials in the proceedings of the Deya conference on "Ritual, Rites and Religion in Prehistory" (Waldren et al. 1995).

Table 4.1. Wealth and Pathologies at Gatas

Categorization by wealth (after Lull and Estévez 1986), 1 = wealthiest, 5 = poorest	Total number of burials	Total number of burials with dietary or disease stress pathologies
1	0	0
2	1	0
3	8	1
4	8	1
5	12	3

reminds us that "however strong a man, he is never strong enough to remain master always unless he transform his Might into Right and Obedience into Duty." The archaeological record for the Argaric presents abundant evidence for might, and perhaps some for obedience, but right and duty barely make an appearance. Coercion (or at least violence) is materialized, but its legitimation is not.

This absence of any evident consecration of the social order contrasts strongly with earlier and later periods of Iberian prehistory. In the Copper Age the construction of megaliths and the deposition in them of ritual grave goods give concrete expression to the ceremonies that united lineages and asserted their claims to the land that ensured their survival. In the Iron Age, a variety of sanctuaries, temples, and cult objects demonstrated the divine sanction of the social order of emergent city states. During the Bronze Age, however, communal institutions had lapsed, but civic ones had not taken their place. It is difficult to accept that rulers powerful enough to ensure their succession would not claim supernatural justification for their actions. It would seem, therefore, that it is not just the proponents of Bronze Age states in southeast Spain who have indulged in complexity inflation (Yoffee 1994) in their interpretations of the Millaran–Argaric sequence.

CONCLUSION

These admonitions should not be interpreted as a manifestation of a renewed methodological pessimism. It is true, of course, that after 20 years of functionalist (sensu lato) research in southeast Spain the various processual accounts of the Millaran–Argaric sequence still have to be judged more on their realism than on their factuality. Distinguishing rulers from leaders makes demands on evidence that archaeological fieldwork in Iberia is only beginning to confront. What this review of the state of research indicates, however, is that the competing explanatory sketches can in fact be tested. Even with the patchy, underreported evidence at hand, we can see that some critical evidence for complexity—settlement hierarchies, full-time craft specialization, supravillage economic organization, or even hereditary elites—is lacking or inadequate. At the same time, the past 20 years of research has tended to confirm the evidence for some degree of agricultural intensification. We can propose, therefore, an alternative scenario (that in turn is testable). We would still argue that the development of incipient Mediterranean polyculture in the Millaran Copper Age led to increased opportunities for differential accumulation of wealth and thereby put a strain on communal institutions, a strain that was palliated by the intensification of megalithic ritual. The Argaric Bronze Age would see the resolution of the tension between intensification and communalism, not by the development of a more stable system of class stratification, but by a descent into internecine strife. Agricultural intensification in the Millaran and Argaric generated conflict, but did not generate a surplus sufficient support an elite that could control that conflict. On this reading, then, the tensions within Millaran society led not to a revolutionary reconstitution of society but to the common ruin of the contending lineages.

ACKNOWLEDGEMENTS

Celia de Jong, Benedicte Gilman, María Isabel Martínez Navarrete, Ignacio Montero, Monica Strauss, Sander van der Leeuw, and Juan Vicent provided useful comments on earlier versions of this paper.

Part **III**

Leaders to Rulers

Leadership in emergent centralized polities does not follow a single pathway. Although such societies share a common foundation in the consolidation of certain kinds of decision making—economy, warfare, religion—they also express diversity in the leadership strategies of the decision makers. The transition to politically centralized society involves a change in the nature of social tinkering. While tinkering continues at all levels of society, including leadership, the tinkering of leaders can have a bigger and more immediate impact on the historical trajectory of the society than that of most other individuals. Leaders experiment with alternative strategies in exercising authority and steering the direction of the society as a whole. Thus, for example, if a chief in one society decides to wage war on a neighboring group, the entire society becomes involved in the effort. The future success of that society is then affected by the outcome of the chief's tinkering.

The chapters in this section provide multiple perspectives on both the variety and patterns of leadership manifested in politically centralized societies. In terms of the trajectory of political centralization discussed in Chapters 5 and 6, Kristiansen and Earle, respectively, pick up where Creamer and Gilman left off. Both are looking at developed, relatively stable centralized societies with leaders who exercise explicit but limited roles as centralized decision makers. In Chapter 7, McAnany goes yet one more step to look at the actual transition from the leadership of chiefs to the rulership of kings.

In Chapter 5, Kristiansen bases his analysis on extensive research on what have been traditionally recognized as chiefdom societies in northern Europe and Scandinavia in the 2nd millenmium BC. What he offers is really a totally new way of interpreting the symbolic and political implications of the material markers of leadership in prehistoric polities. In doing so, he illustrates how decision making does not have to be concentrated in the hands of a single individual, but that different individuals may be responsible for making different kinds of decisions.

In Chapter 6, Earle goes to a very different kind of material manifestation of leadership—the very landscape on which these societies existed. Making continued effective use of his research on chiefdoms in Denmark, highland Peru, and Hawaii, Earle shows how cultural exploitation and manipulation of the natural landscape reflects variation in leadership strategies. Earle also shows how staple versus wealth finance economies combine with network versus corporate organizational strategies to give a much more complex and richer range to the kinds of leadership found in centralized polities.

In Chapter 7, McAnany is the first who actually focuses on the transition from leaders to rulers. In this case, the transition is from Mayan shamans with the decision making role of leaders in chiefdoms to Mayan kings with the greatly increased authority and power of rulers in state societies. She has conducted extensive field research at the Mayan center of Naj Tunich, which she uses as the foundation for her analysis of the relationships between principles of kinship and the rule of Mayan kings. In Chapter 7, McAnany extends out to look at Mayan society in general and show how the relatively informal decision-making role of shaman leaders comes to be transformed into the institutionalized role of kings.

Chapter **5**

Rulers and Warriors
Symbolic Transmission and Social Transformation in Bronze Age Europe

KRISTIAN KRISTIANSEN

SYMBOLIC STRUCTURES AND SOCIAL INSTITUTIONS

Symbols gain meaning through context. Contextualized meanings are of many kinds, some of which are not available to archaeologists. Myth, dances, and other types of performance are the primary contexts of symbolic meaning, only rarely available to archaeology. They represent the narratives that give life to social and ritual institutions. But myth and narratives are also materialized, in paintings, decorations, consumptuary goods, ritual structures, and so forth, which often are bound together through common rituals. In the past such different material remains would have been understood as being part of common ritual events and myths, while in archaeology they are separated and defined as different contexts, labeled burials, standing stones, rock art, and votive offerings/sanctuaries. Since archaeological categories are normally the starting point for research, from the beginning we are being constrained in understanding what possible symbolic and cosmological meanings and what possible rituals and myths unified such different

---

KRISTIAN KRISTIANSEN • Department of Archaeology, Gothenburg University, Gothenburg, Sweden SE 405 30

From Leaders to Rulers, edited by Jonathan Haas. Kluwer Academic/Plenum Publishers, New York, 2001.

remains. To rediscover these lost intercontextual meanings demands an integrated, holistic approach. I believe we are now in a position to take such an interpretative step by linking symbolic structures to social institutions. To do so we need the integrating concepts of materialization and institutions, emulation and permutation.

Institutions are the building blocks of society. To trace them therefore is a central interpretative task. Institutions are also invisible, but they often materialize in specific and recurring ways that allow one to infer the cultural and institutional significance of the evidence (DeMarraiset et al. 1996; Earle 1997: Chapter 5). The interpretative task is to single out among the different objects and contexts available to us those that formed a specific relationship defined by a set of symbolic meanings, actions, and transactions that once linked them together in an institution. Tracing the origin of institutions, whether in time or space, has to be done for each region/period under study, so that we are able to compare recurring contextual relationships that define institutional relationships. These may be movable in the form of certain types of objects, or immovable in the form of certain types of monuments, architecture, paintings, ceremonies, and so on. The problem or the challenge is that foreign material evidence is often translated into the local cultural language, which means that only bits and pieces of the original evidence survive. It is acculturated and recontextualized. In this process the original meaning and message often is carried on in a selective way leading to the adoption of new value systems, practices, and eventually institutions to reinforce them. If that is the case it is only by interpreting and comparing institutions that we are able to understand and eventually explain the true historical impact of social and economic interaction between societies. This will be exemplified in what follows.[1]

Emulation and permutation are formal properties that characterize certain central symbols that can be applied in a variety of contexts. Permutation allows for variations in the order a certain symbol is used. Thus although the original order of things is unknown to us, we may be able to understand symbolic connections irrespective of that order. Emulation means imitating or applying strong symbols or even institutions in another context. Both symbolic and institutional imitations are well known to us. Any Christian recognizes the cross as a basic religious symbol, with a variety of mean-

---

[1] In that process we are confronted with the difficult task of analyzing and understanding the dynamic relations between agency and convention, event and structure. To do that we must attempt to trace the socially and culturally determined motives and incentives for individuals to travel and to adopt new values and behaviors. It is a difficult task that cannot be done by applying twentieth-century urban experience. It demands systematic studies, of the rationalities and motives behind travels and journeys in prestate and early state societies (Helms 1988, 1993, Kristiansen 1998a).

RULERS AND WARRIORS

ings depending on its context, from being a general symbol of Christianity, being worn as ornaments by women, in warfare by the crusaders, as a burial cross, as the ground plan in church architecture, as a crucifix for prayer and sacrifice, and so forth. The pervasive and emulating nature of the Christian cross is common to all central religious and cosmological symbols. I am suggesting that by tracing symbols through their different contexts, it will be possible to detect recurring and meaningful mythological and cosmological structures in the archaeological record, if they existed, as demonstrated by Hodder's (1984) study of Neolithic houses and burials. When textual and mythological evidence can be added, new historical and religious meaning will suddenly emerge behind traditional archaeological typologies and classifications of art and rituals, as demonstrated by Hedeager (1997) for the Iron Age, Shanks (1992) for Archaic Greece, tracing the meaning of a perfume jar, and Kaul (1998) for ships on bronzes. Figure 5.1 exemplifies two interlinked interpretative strategies. However, we have no guarantee that specific strong symbols will always be connected by intercontextual

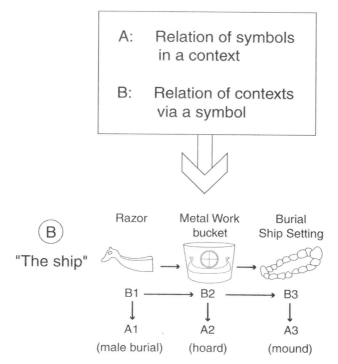

Figure 5.1. Interpretative strategies for tracing the meaning of symbolic contexts and institutions.

meaning; it remains to be tested in each case. But I do suggest that it represents a powerful interpretative strategy.

In the following I shall demonstrate the usefulness of such an approach on the Nordic Bronze Age, so rich in symbolic evidence. I wish to stress that I am not denouncing the importance of studying specific archaeological contexts such as burials or hoards, since there is still much that needs reinterpretation, just as many neglected features from excavations can be interpreted as possible ritual representations (Kaliff 1998). However, my goal here is to demonstrate that it is possible to become more specific about religious and cosmological meanings through a specific theoretical strategy of interpretation.

THE TWIN RULERS: THE RITUALIZED STRUCTURE OF POLITICAL LEADERSHIP

Tracing the Meaning of a Symbol: The Chiefs Cap

We shall encircle the cosmological universe of the Nordic ruler, or rulers, by tracing a single chiefly emblem: the cap and its symbolic applications through various contexts. As is well known from Near Eastern contexts, hats were a special emblem of rulership, and their significance in the European Bronze Age has recently been demonstrated by Randsborg (1993:111 ff) and Gerloff (1995) in an original interpretation of the golden caps of western Europe. Among the Hittites it could be both tall and pointed or rounded, the latter belonging to kings and the first to gods (Larsson 1997:Figure 5). The rounded version is dominant in the Nordic Culture of Montelius II (1500–1300 BC), where it has been preserved with textiles and clothing in several of the oak coffin burials (Broholm and Hald 1940). The Nordic cap comes in two versions: a complex high-quality version covered by a close pile consisting of fine, short threads ending in a knot. These are really small technical masterpieces and have been time consuming to produce (Figure 5.2A). An additional simpler version without pile are found in some coffins, such as Trindhøj or Guldhøj (Figure 5.2B).

In a group of rich chiefly burials, small hat–formed bronze tutulus in pairs, probably used for the sword strap (Poulsen 1983:Fig. 1, Tables I–III). They may symbolize the round cap and its bearer (Figure 5.3C). The same symbol is applied on the richly spiral-decorated cult axes that are always found in pairs in hoards. The axe would thus symbolize two male persons

Figure 5.2. (A) Elaborate cap from Muldbjerg, Jutland (after Broholm and Hald 1940:Fig. 12). (B) The mans dress in Trindhøj, Jutland, with a taller and simpler cap (after Broholm and Hald 1940, Fig.35).

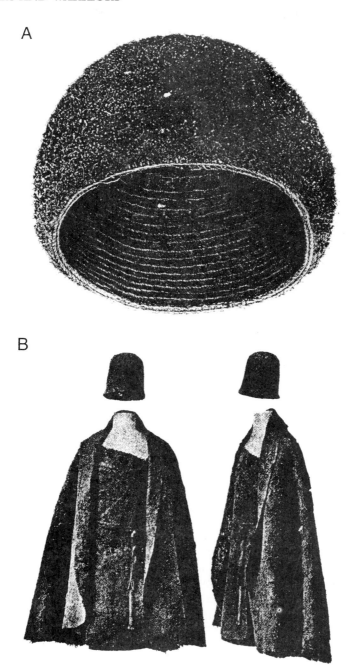

A

B

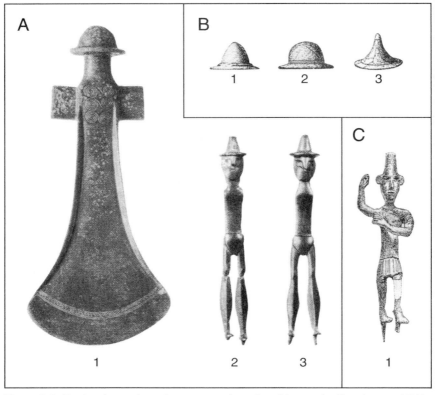

Figure 5.3. Tracing the cap in various contexts from the mid-second millennium BC. (A) Two male figures with tall hats and an axe god with rounded hat from the Stockhult hoard. (B) Hat-formed tutuli from chiefly burials. (C) Imported levantine figurine with a tall hat found in Lithuania, probably a thunder god holding a spear (after Gerloff 1995:Abb. 9).

(gods) with rounded caps on (Figure 5.3). We meet these axes in cultic processions on rock art, often carried by two males (Figure 5.4). But the twin cap bearers also materialize as bronze figures in the famous Stockhult hoard. Here the cap is rather tall, reminiscent more of the Hittite type, but also resembling the other cap in the Trindhøj burials, which was tall (ca. 22 cm).[2] In tracing the ruling symbol of the rounded cap we have been able to

---

[2] Note the loincloth of the two warrior gods, in style similar to Minoan and Hittite war gods, or "young warriors" (Marinatos 1995:Tf XV a-b, XVIb), but unknown in Scandinavia. It raises the question of whether the figurines have an East Mediterannean origin. Also the movable, now missing arms represent a foreign trait, unique to the figurines in Scandinavia. I therefore conclude that they were East Meditereannean imports or direct imitations of imports.

RULERS AND WARRIORS

define a specific chiefly institution: the chief who also is the ritual leader of axe processions in rituals. We further have gotten an indication that this institution consisted of twin chiefs, at least in certain ritual contexts. Would the twin leadership apply more widely and can they be documented and defined more precisely in other contexts?

Tracing the Contexts of the Twin: Dualism

Having now surveyed the chiefly institution of the twin hat and axe bearers, we shall search for contexts that may support and add meaning to it. As we saw above, the twin axes with the round cap are found deposited

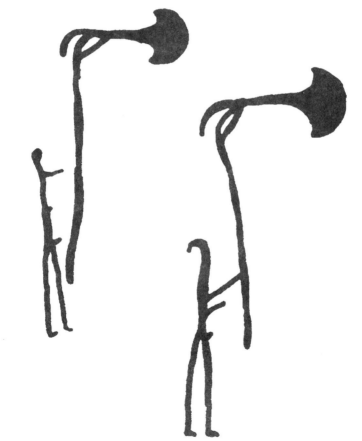

Figure 5.4. Rock art from Simris of the twin cult axes carried by two males (after Kaul 1998:Fig. 6)

pairwise in hoards, never in burials. They are represented on many rock carvings in processions, often in pairs. We may here assume a relationship in meaning between the pairwise deposition and its employment in rituals. As they never occur in burials, these axes belonged in a religious and ritual sphere that could only be performed and taken care of by a priest–chiefly priest who would sometimes deposit the axes as a gift to the gods. Later in the Bronze Age the twin chiefs with axes appeared as miniatures in the famous Grevensvænge finds (Kristiansen 1998a:Fig. 47), and here we can observe that the ritual twin chiefs/gods are wearing a special dress. The front ends in a flap below the belt and on the back it has the form of an animal tail, perhaps symbolizing the tail of the ox as represented by the horned helmets. However, in an early Period III burial find of a Bronze Age chief with sword, razor, tweezer and knife, a piece of cloth was found in the form of a flap (Kaul 1998; Figures 2–5). Furthermore, this chief also had a

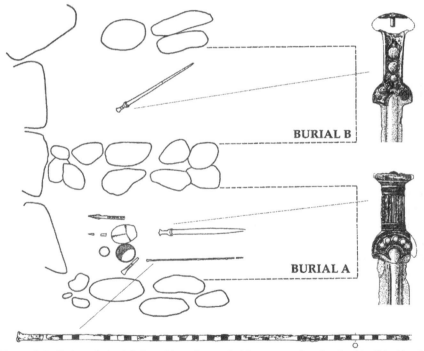

Figure 5.5. Twin male burials from Montelius period 2, representing the ritual/political leader with Nordic full hilted sword, a staff of bronze and a full equipment, including razor/tweezers and a drinking cup, and alongside him the warrior chief with a foreign flange-hilted sword as only grave goods (art work Thomas Larsson, based on Aner and Kersten, Band 4, 1978:Abb. 202 and Tafel 76–77)

belt purse full of amulets and had the razor/tweezer. Furthermore, the burned bones of a female suggest the chief was buried with a human sacrifice, a phenomenon not uncommon in the chiefly burials.

Here we then have a ritual chief and priest with special magical equipment and special dress, suggesting magical and even shamanistic functions of taking on animal guise, something we also see on many rock carving of ritual scenes. The chiefly priest, however, was also a war chief, since he normally had the full-hilted chiefly sword and a war axe, very different from the ritual axes. But as I have demonstrated earlier (Kristiansen 1984), the full-hilted swords were not primary for use in warfare; rather they were a ceremonial weapon, in opposition to the functional flange-hilted swords.

Thus we have defined quite clearly the joint chiefly and priestly roles of the leader. But who was the twin? It could be another similar chief from a neighboring community, or it could be within the family so to speak. Specifically, it may have been the warrior chief with the flange-hilted sword, who never or rarely would be buried with items of ritual or other social chiefly functions. A double or twin male burial from southern Jutland confirms the pairing of priestly and warrior chiefs (Figure 5.5). Here we find side by side the priestly high chief with the chiefly Nordic full-hilted sword along with his "twin" ruler, the warrior chief with his "foreign" flange-hilted sword. The high chief even has a ruling staff of bronze, as we know them from Minoan Crete (Marinatos 1995:Pl.XV–XVI). This rather unique double burial confirms a special relationship between the ritual chief and the warrior chief, a dual political/ritual institution of leadership. Normally the chiefs were buried individually, but always under a chiefly barrow.

If we again return to the chiefly cap and dress: the warrior chief also had the special rounded cap. In that he is unified with the ritual chief as a person of chiefly rank. Both were wearing a large cape that in the Iron Age would characterize the officers in warfare.

Due to the exceptional conditions of preservation in Danish Bronze Age oak coffin burials we can for the first time in European history define elements of a chiefly costume consisting of cape and the cap of the ruler. I believe these two elements of dress were accompanying the introduction of the institution of warrior aristocracies in the East Mediterranean and central Europe. To date, they can only be documented in the Nordic Bronze Age.

We still do not know how integrated the twin chiefly functions were in daily life. Insights, however, are provided by a series of remarkable settlement excavations from recent years of Bronze Age farms, among them chiefly halls of sometimes incredible size: 30–40 m long and 8 m wide is the typical chiefly farm hall of this period, which appears together with the new institution of warrior aristocracies and twin rulers. In these big three-aisled halls we find a bipartite architecture (Figure 5.6): There are two identical living

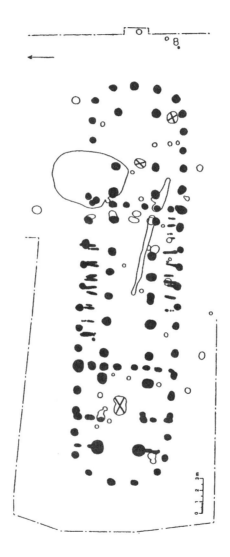

Figure 5.6. Large farm building from Montelius Period 2 with two identical living quarters and stalling for cattle in the central part. From Legård, Thy (after Kristiansen 1998c).

RULERS AND WARRIORS

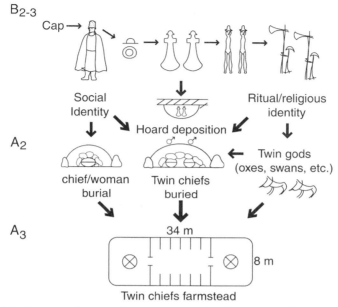

Figure 5.7. Summary diagram of the strategy employed to support the interpretation of the institution of twin rulers in Bronze Age Scandinavia.

quarters each with a hearth. In the house on Figure 5.6 we find the central hall occupied by the most costly prestige good: stalls for the cattle. In other farm halls there are no traces of stalls. The house has two entrances, on each side of the house close to the living quarter. So this is a farm hall for two families with their cattle. Even then they had nearly 100 m^2 in the living quarter, which was further subdivided, a truly chiefly compartment.[3] This bipartite structure is a recurring feature of the larger halls in the Nordic Bronze Age, and we may therefore assume that the twin chiefs were as integrated in the social and religious organization of society as indeed suggested by the rock art and by the pervading twin symbolism in material culture. Figure 5.7 summarizes the interpretative strategy supporting the reconstruction of the institution of twin rulers—chief of rituals and chief of war—according to the interpretative model of Figure 1.

[3] A farm hall of identical size and contsruction was lying parallel to the one on Figure 6. The excavators, including the author, believe that this "twin hall" was erected after the building for the first house (I), since the post holes are smaller, suggesting a beginning scarcity of wood and at least a slightly more light construction than the rather massive timber posts in house I. But it cannot escape notice that this house was build as an exact twin to house I, and placed parallel to it.

Was this a genuine Nordic tradition or was it adopted along with the new institution of warrior aristocracies (Kristiansen 1999). As I have demonstrated elsewhere, the rise of the Nordic Bronze Age and at the same time of chiefly society was due to a tremendous creative and organizational capacity, transforming external influences into a new genuine cultural and social system with but few antecedents on home ground (Kristiansen 1987, 1998a). By tracing structures of meaning behind symbols of rulership and through contexts, as prescribed in our theoretical introduction, it has been possible to go behind the stylistic patterns and detect a structural pattern constituting the chiefly institution of twin rulers. We shall now briefly trace its origin.

Origins of the Twin Rulers

As already stated, the institution of twin rulers as it unfolded from Montelius I into Period II was linked to the adaptation of the warrior aristocracies during the same period. We shall now demonstrate some of the evidence supporting this proposition. But before doing so we should examine whether an older tradition of twin rulers existed in Scandinavia. We do find slight indications that a social and religious tradition of double male burials existed going back to the Single Grave/Battle Axe culture, which was not integrated, however, into an institution with its own symbolic language and which occurs very rarely. Glob (1944:179) mentions a handful of examples from the hundreds of burials analyzed in his work. The development of the institution of twin rulers, however, could have been facilitated by this existing tradition.[4] We get another contribution to the history of twin rulers from Indo-European religion. Here a recent study has revealed that twin rulers were a consistent structural feature of proto-Indo-European society but were to disappear in later Indo-European religion (Olmsted 1994). However these aspects are beyond the scope of this chapter. Instead I refer to the forthcoming book (Kristiansen and Larsson, in press).

It is by the beginning of the Bronze Age proper in Scandinavia (1700 BC) that the twin symbolism is introduced, linked to the double axe symbolism. Vandkilde (1996) demonstrated by this time the emergence of a class of high-quality axes that were not used as axes but rather in rituals and as a symbol of rank and that were regularly deposited in pairs and now also appeared on rock art in rituals. From 1500 BC the institution of the twin rulers was a penetrating phenomenon, which continued into the late Bronze Age, where it also came to include female depositions (double neck rings,

---

[4]More systematic research is needed to substantiate this proposition, which falls beyond the scope of our work. But today thousands of burials are at hand from the corded ware and Single Grave cultures in Europe, so it should be possible to answer the question: Did there exist a late Neolithic tradition of male twin burials?

double sets of ornaments in hoards). During the Late Bronze Age ritual paraphernalia of the twin rulers reached a high peak, as it was now accompanied by bronze shields, helmets, and lurs, which often are found deposited in pairs, and was used by twin chiefs/priests in the rituals as amply demonstrated on rock art and on Bronze figurines, for example, Grevensvænge. But during the early Iron Age the twin symbolism disappeared. So in terms of origin, one should look for evidence from the period 1700 to 1500 BC.

Here we can point to the Minoan–Carpathian connections (Kristiansen and Larsson, in press). Among the many influences that were channeled northward were the ritual institution of the axe gods.

The closest parallels to the twin gods and twin rulers are to be found in the Minoan and later Mycenaean culture where it was probably adopted in the later so-called *wanax* institution of the king. As Kilian (1988) has argued, the royal megaron was bipartite; and Palaima (1995) in a recent exhaustive discussion of the texts on the *wanax* suggests that it was taken over from Crete, together with other institutions. He further makes clear that the *wanax* main duties were religious, although he could also have some concern for warfare. In that he is like the Hittite king (Klengel 1996). The textual evidence makes clear, however, that there is a special chiefly or royal commander of war, named *lawagetas*, who was second in power after the *wanax*. What do we have here, other than the twin rulers, as we see them in the North? If Palaima is correct, then it has two implications: It seems reasonable to conclude that Mycenaean ritual adopted not only Minoan religious symbols and status symbols in the shaft grave period (bull heads rhyton, double axes, horns of consecration) but also in a selective way institutional traits, such as the institution of the double axe and the twin gods linked to them (Hägg 1984). Otherwise it is difficult to understand how the *wanax* and other Minoan institutions should appear so developed by the palace period. It is the combination of axes in rituals and sometimes bull horns (found on one ritual axe in Denmark) that was transmitted via the Carpathian tell cultures, although we find no explicit twin symbolism here. But we find axe cult and numerous hoards with ritual axes, some of them with pairs of axes, a few with bull horns, but most, however, with the conical hat symbol. There is no clear distinction between war axes and ceremonial axes, as in Scandinavia, but in several hoards we see the combination of chiefly full-hilted swords and richly decorated axes, as in the famous Hajdu Samson hoard.

Second, it implies that the *wanax* structure with its division of power between priest king and warrior king should be found on Crete. The camp stool frieze on Knossos with the twin gods/young males sitting on camp stools face to face lifting the cup to cheer is the natural point of departure for understanding the institution of twin gods in Minoan culture. In a recent discussion, Rehak (1995) referred to a reconstruction of the frieze made by Cameron (1967) that links it to a chariot scene, with on older man in robe with diagonal band who leads a tethered bull. The female priestess or goddess on the throne is presiding over the twin males and the whole ceremony. If these additional elements are accepted, we can link the ceremony to some form of initiation or ritual of young chariot warriors of high rank. Twin gods/goddesses also are driving the chariots on the Haiga Triada fresco. These scenes suggest an institutionalized structure of twin warriors/warrior gods in Minoan society, but with priestesses presiding over ceremonies and sometimes even driving the chariot. The drinking scene has much in common with the drinking ceremonies of the chiefly or royal retinue as it is known in the Iron Age of Europe, where a priestess or high-ranking women plays an important role (Enright 1996), but a closer parallel is found on the Hittite relief-vase from 1600 BC with a complete ritual sequence, including libation, bull slaughtering, and final drinking. Also here we find twin figures and the final drinking scene is two sitting persons in front of each other, one on a camp stool lifting his glass (Klengel 1996:Abb.3; Randsborg 1993: Fig.70). Fragments of a similar scene is found in the processional frieze in the vestibule and the throne room within the Pylos megaron complex, stressing the ritual and religious uniformity of the Bronze Age cultures of the East Mediterranean (Palaima 1995:Pl.XLIa-b).

The Minoan evidence and the textual evidence on *wanax* and *lawagetas* add significance to the institution of twin rulers in Scandinavia and suggest that we are speaking of the same institution adopted to different social environments. When we add to this the evidence of the axe gods/bull horned god that we find from Minoan Crete over the Carpathian to Scandinavia during the same period and whose Minoan iconography even was adopted in the Kivik burial, then it seems safe to conclude that a complex transmission of religious and political institutions took place during this period from Minoan Crete to Nordic Scandinavia. Not only chariots but even the camp stool accompanied this transmission, as it was generally employed in chiefly burials in southern Scandinavia. The new institution of twin rulers was accompanied by the formation of warrior aristocracies and retinues throughout Europe, which we shall briefly describe.

THE SOCIAL AND CULTURAL CONTEXT OF WARRIOR ARISTOCRACIES

From the eighteenth–nineteenth century BC two interlinked phenomena spread across Europe: a new weapon complex that employed long

RULERS AND WARRIORS

sword, lance, and chariot. It represented new military tactics, originating in the empires and palace cultures of the Near East and Eastern Mediterranean, based on the employment of chariots to supplement infantry. The new weapons meant more heavy man-to-man fighting and demanded new military skills and the employment of protective armor. It thus put new demands on the training of warriors and subsequently on their social and economic support. The professional warrior, well trained and organized, was introduced.

In temperate Europe the new weapons were linked to the rise and expansion of a new aristocratic warrior elite, above the traditional tribal warrior, which had employed bow/arrow and dagger/war axe since the 3rd millennium BC or even earlier (Kristiansen 1987). In the archaeological record the new warrior aristocracies set themselves apart by being buried in richly furnished burials, often in a barrow with sets of weapons, which also could be deposited in hoards. From the sixteenth–fifteenth century BC, the new warrior chiefs became a common phenomenon at both local and regional levels in temperate Europe; social and military differentiation was recognizable in different combinations of weapons in grave goods and in different uses of weapons (Coombs 1975; Kristiansen 1984, 1987; Schauer 1984, 1990).

This pattern of chiefly war leaders with a retinue of lance warriors and a smaller group of chiefly warriors is seen to be consistent over wide regions from the sixteenth century onward (Figure 5.8). Functional changes in the nature of combat and the preference of weapons occur through time

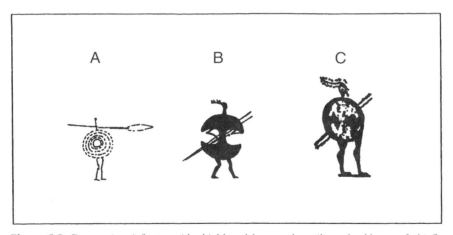

Figure 5.8. Bronze Age infantry with shield and lances, the military backbone of chiefly retinues during the second and first millennium BC. (A) Nordic warrior from the Wismar bronze horn (Middle to Late Bronze Age, Montelius period II/III). (B) Mycenaean warrior fro an amphora. (C) Greek warrior from a Geometric krater.

and were quickly adopted from the Mediterranean to northern Europe, such as the introduction of the efficient flange-hilted warrior sword from around 1500 BC. During the early 2nd millennium BC, the lance and the short sword/dagger were dominant, to be replaced by the long sword from the seventeenth century BC. A shift from rapier/axe combinations to slashing swords/knife took place from the late fourteenth century BC, while after 1000 BC lances again gain dominance as the standard weapon of the hoplite, as well as in central and northern Europe. Numbers in battle are difficult to estimate, but they probably varied from small raiding parties, reflected in small weapon hoards, to armies numbering in the hundreds attacking the fortified settlements, as suggested by both Randsborg's (1995:44ff.) and my own studies (Kristiansen 1998a).[5]

The appearance of warrior aristocracies represent the formation of a new chiefly elite culture in Europe (Kristiansen 1987, 1999; Treherne 1995). It was embedded in new rituals, in new ideas of social behavior and lifestyle (body care, clothing, etc.), and in a new architecture of housing and landscape (Kristiansen 1998c). It centered around values and rituals of heroic warfare, power, and honor and it was surrounded by a set of new ceremonies and practices. They included ritualized drinking,[6] the employment of trumpets or lurs in warfare and ritual, special dress, special stools, and sometimes chariots (Figure 5.9). It meant that chiefs were both ritual leaders and leaders of war as demonstrated above.

Thus, the new chiefly elite culture spread as a cultural package, a new social value system, rather than as separate elements. We may characterize it as a new social institution of chiefly leadership that integrated ritual, rank, and coercion. It therefore became a penetrating phenomenon, crossing cultural boundaries throughout Europe. Although it was adapted to local or regional cultural traditions, its main components are easily recognizable from the Mediterranean to northern Europe. And since appearance was an

[5] Robert Drews in a recent book has suggested that the military organization of the central european Late Bronze Age infantry was superior to East Mediterannean armies employing chariots and small groups of foot soldiers. The "barbarian" armies therefore were able to overrun the more traditional organization of the East Mediterannen armies during the turbulent thirteenth–twelfth century BC, which changed both the balance of power and military organization (Drews 1993)

[6] In a most interesting book, *Lady with a Mead Cup*, Michael Enright (1996) has demonstrated continuity in the drinking rituals and social reproduction of chiefly retinues from La Tene to Viking Age, based on literary evidence but also supported by archaeology. His findings, however, are echoed in the material culture of the preceding millennium. From the early Bronze Age of Scandinavia we find cups with relics of mead in both male and female burials (among them the famous Egtved woman wearing a corded skirt employed in rituals, as demontrated in female figurines, just as one figurine holds a cup). Drinking sets for several persons in bronze and gold are likewise associated with both male and female depositions during the Late Bronze Age (Kristiansen 1984:Figs. 10, 11).

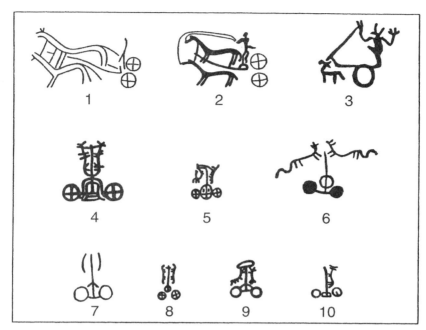

Figure 5.9. Chariots displayed on rock art in Scandinavia during the sixteenth–fourteenth centuries BC. Note the very accurate representation of the basic construction of the chariot as seen from above on No. 4.

important factor, it left rather good archaeological traces as demonstrated above. I shall now proceed one step further and propose to characterize the social and cultural institution of warrior aristocracies. It comprises the following interlinked elements:

- New body culture and dress codes. Appearance and body care gained new significance, reflected in the universal employment of razor and tweezers, and in needles for tattooing (Treherne 1995, Willroth 1997:Abb. 6). Men were wearing elaborate hats and capes, women were wearing complex hairstyles and elaborate ornaments, sometimes constricting their physical movements (Sørensen 1997). The chiefs were sitting on stools and drinking mead from costly metal vessels and amphorae, or more often, from imitations made of wood, but decorated sometimes with tin nails to signal status.

- A new architecture was introduced in the form of large chiefly halls/farms that could hold the chiefly lineage and attendants, as well as also sometimes the cattle. In western and northern Europe the landscape was reorganized, with chiefly barrows placed on all hilltops, surrounded by grazing cattle/sheep (Kristiansen 1998c). In central Europe fortified chiefly residences and villages, sometimes with an acropolis emerged (Jockenhövel 1990). Specialist metalworkers were attached to the new residences.

- New warrior lifestyle. The new forms of combat demanded regular training, as well as specialists to produce weapons and wagons. Thus a new warrior lifestyle became an integrated part of traditional life at every local chiefly farm. In opposition to a state society with a professional army separated from daily life, daily life was dominated by warrior values in addition to harvesting the fields and raising the cattle. Warrior ideology and lifestyle were ingredients in everyday life and the meaning of life to most young chiefly sons. Since every parish had its local chief, sometimes several, the influence of the chiefly warrior culture was penetrating social life.
- New social organization of warfare. This was reflected in the new system of clients/retinues, which was the basis for mobilizing war parties for raids, trading expeditions, and so forth. But it also was the economic basis for the chiefs, since clientship allowed the collecting of surplus to finance feasts, boats for trading expeditions, rituals, and warfare. While the chiefly barrows and feasts reinforced chiefly power and generosity, the professionally trained warriors were the means to extract tribute from unwilling clients and to enlarge tribute through raids when needed.

Thus, in the Bronze Age and early Iron Age of central and northern Europe warfare was an integrated aspect of daily life. Every local community would regularly bury their dead warriors. Farming and warfare were the two axes of male activity, war training being a main concern of the young local chiefly males. During seasons of war local retinues were mobilized along lines of rank. Changes occurred in sword combat: During the earlier 2nd millennium the long and narrow rapier was dominant (thrusting), while from the thirteenth century BC the wide bladed and heavier slashing sword took over. The size of armies may have developed over time, and thereby increased the scale of combat and political control, but this is mainly based on the size of fortifications (Kristiansen 1998a:Fig.200), whereas the overall social organization of warfare based on the chiefly retinue remained largely unchanged. The impact of warfare on Bronze Age populations has never been studied systematically but can sometimes be demonstrated in the age statistics of cemeteries. Here we find a recurring gender variation: young women (juv./juv.adult) show higher mortality than men owing to the risks of childbirth, while adult men demonstrate highly increased mortality compared to women, most likely owing to warfare (Kristiansen 1998a:Fig. 198; O´Shea 1996:Fig. 6.1). In some cases we also find the victims of warfare quickly buried or sometimes even thrown into mass graves, examples numbering from a handful to several dozens (Louwe Kooijmans 1998:Abb.12; Chochorowski 1993:Figs. 47, 48).

RULERS AND WARRIORS

In all this the Bronze Age and early Iron Age societies of the 2nd and 1st millennium BC resemble historically known chiefdoms, which were characterized by systematic warfare, including a rather high proportion of the male population and many casualties (data in Keeley 1996:Figs. 4.1, 6.1, and Tables 2.1,2.6, 3.2, and 6.2).

Although this social system entailed many of the building blocks necessary for state formation, such as chiefly retinues and the extraction of tribute, the process was constantly being held back by internal dynamics and constraints. Social dynamics were mainly linked to the ritualization of power in combination with the competitive nature of exchange, whereas constraints were ecological and demographic.

Chiefly warrior aristocracies and warrior culture consequently remained an inherent feature of the social and ideological organization of European Bronze Age and early Iron Age societies throughout 2000 years, probably 3000 years. We thus have to project back in time the old notion of *Militärische Demokratie*, as coined by Morgan and later absorbed by Engels, with another 2000 years.[7] It raises a series of questions as to the historical role of warrior aristocracies in the evolution of early states, characterized by a decentralized political economy. I suggest that this social form was one of *longue durée*, lasting for several thousand years, not as a cause but as an effect of its structural position in a larger world system in combination with environmental constraints.

SYMBOLIC TRANSMISSION AND SOCIAL TRANSFORMATION IN BRONZE AGE EUROPE

In this chapter I have demonstrated the role of institutional borrowing in Bronze Age Europe, exemplified in the symbolic transmission of its constituting symbols. While material culture may adapt to different cultural contexts or "dialects" (style groups), the symbolic structure of the dominant institutions remained intact. The spatial and temporal history of two interlinked institutions was traced: the twin rulers and warrior aristocracies.

Their expansion corresponds to a period of social transformation in Europe due to a tremendous interaction between the societies from the East Mediterranean to Scandinavia, the impact of which has been underestimated. By applying a new theoretical understanding of the relationship between materialization and social institutions and a new interpretative strat-

---

[7] I am not discussing the vast literature on *Militärische Demokratie* or the Germanic mode of production (see Herrman 1982; Gilman 1995a), since I believe that we need more up-to date case studies to establish a better empirical and historical platform for discussion

egy, it has been possible to exemplify the nature and the social and religious implications of this interaction. It further demonstrated that religious and social institutions of leadership were integrated in Bronze Age Europe. Thus ritual, rank, and coercion were inseparable aspects in the expansion of ruling elites.

Chapter **6**

Institutionalization of Chiefdoms
Why Landscapes Are Built

Timothy K. Earle

INTRODUCTION

Our challenge is to understand the multilinear strands of social evolution that emerge within constraints and opportunities of local histories and environments (Trigger 1998). Although all human cultures are intricate and complicated in different ways, we can focus attention on the variables related to the development of political organization: "A consistent element in the broad course of cultural evolution has been the emergence and subsequent development of centralized forms of political organization" (Haas and Stanish 1996). As the size of polities have expanded, new levels of integration have been created to centralize, or at least syncretize, human political action. Elsewhere I have identified the interlocking economic, political, and social reasons why complex societies develop (Johnson and Earle 1987), and I have emphasized how the control of power is critical to centralization (Earle 1997). Power derives from control over material flows through the human

---

TIMOTHY K. EARLE • Department of Anthropology, Nortwestern University, Evanston, Illinois 60208.

From Leaders to Rulers, edited by Jonathan Haas. Kluwer Academic/Plenum Publishers, New York, 2001.

economy, and this control rests, at least in part, on the ability (might) to exclude others from desired things.

Starting from a materialist perspective, we must understand how the infrastructural conditions of environment, population, and economy determine political structure. For the evolution of political complexity, the major step is to establish systems of property through which rights in productive resources are held by a ruling elite (Earle 1991c, 1998). Such power to exclude rests ultimately on military force, but also on institutionalized rights of access that are codified in a society's ideology and materialized by concrete symbols from copper plaques displayed in ceremonies, to monuments in the landscape, to legal documents enshrined in bureaucratic archives (DeMarrais et al. 1996).

What changes as human societies become increasingly complex? That story has been told many times before, and among those who accept social evolution, a basic consensus about its plot exists. I want to highlight, however, what I consider to be a poorly understood aspect of social evolution: namely, *institutionalization*. Social institutions are human organizations formed for particular reasons. Humans build institutions that integrate more people in increasingly differentiated ways, but especially they determine rights and compensations within increasingly complex economies.

British social anthropologists have emphasized *social structure* as the form for human institutions. Social structures are "all social relationships of persons to persons" and "the differentiation of individuals and of classes by their social role;" they are "the concrete reality . . . which links together . . . human beings" and which "are just as real as are individual organisms" (Radcliffe-Brown 1952:190–192). Social structures formalize relationships among people, synchronize human activities across space, and establish continuity through time. But what is the concrete in Radcliffe-Brown's realities? What is the essence of a social structure? We must establish the media from which social structures can be fashioned. Unlike other animals, human social organization is only minimally encoded biologically. Evolutionary ecologists argue that cooperation and reciprocal exchanges within small-scale foraging societies can be traced to our human nature fashioned by natural selection (Winterhalder 1997). Certainly humans are social animals and many aspects of human organization may have biological roots, but the diversity of human societies (especially in the scale and complexity of their institutions) requires a more elaborate model. Social structures are part of culture and often have been seen as norms of human behavior shared among a society's members and passed down across generations. Although such notions of routine interaction may help to understand small-scale social institutions, they are inadequate to understand the complexity inherent in the structured relationships of chiefdoms and states.

INSTITUTIONALIZATION OF CHIEFDOMS

For complex societies, to provide continuity in space and time social structures must exist in forms outside individual minds.

In a consideration of culture and ideology, DeMarrais et al. (1996) emphasize that various media exist that materialize culture external to the human brain. These media include ceremonies, symbolic objects, landscapes, and writing; they give culture physical forms that can be experienced broadly and thus shared to some extent. In modern state societies, institutions are materialized as legal entities incorporated with written contracts. Following on Maine (1870), we might expect institutions in nonliterate societies to be based on kinship, and they certainly are; but kinship alone would seem inadequate to construct the large and formal institutions of chiefdoms and states. Looking to the other media for materialization, ceremonies provide scripts for group interaction and symbolic objects provide vivid means to objectify relationships, but landscapes are an especially effective medium for institutional formation.

Landscapes provide a particularly good medium from which to construct social institutions, because they furnish scale (ability to be experienced by a large group of people), exposure (daily experience), and permanence (stability across time). Natural landscapes often are used culturally to represent a group's organization. Australian Aborigines describe their group's histories in terms of distinctive features in their desert environment such as rocks, rock shelters, water holes, large trees, and the like. Important historical or mythic events of each group are located in relation to these features. They often are marked by elaborate rock art that makes their cultural meaning more permanently associated with individual groups, and at such sites ceremonies are performed that define the group and reinforce its social and historic charter (Sutton 1988; Rapoport 1994). Landscapes are not static; they are transformed by human use and work. Complex societies build many facilities, such as houses, fields, irrigation canals, trails, paths, dance grounds, burials, and monuments, that transform the environment into a cultural artifact. Construction of the cultural landscape is a means to build large human institutions in the processes of social evolution.

The role of landscapes for institutions offers an exceptional opportunity to social archaeology. Archaeologists are often characterized as disabled social scientists; as anthropologists, we are dedicated to studying culture and society, but we can be seen as lacking most evidence for anything but simple and derivative descriptions of human material culture. If we focus on landscapes, for example, some argue that it is memory, meaning, and the sense of place that are central (see, for example, Thomas 1991), and archaeologists are poorly equipped to study these aspects of the human landscape. But if we shift instead to consider how institutions are materialized in the landscape, then the nature of our evidence appears robust.

The cultural landscape is basic for institutions for two reasons: (1) permanence gives stability through time; and (2) the simplicity of understanding transfers across diverse cultural viewpoints. First, permanence is the nature of many constructions. Stone houses, village ramparts, boundary walls, and ancient trails can be described by archaeologists, because their physical mass makes them preserve well. But what about the archaeologist's dilemma of missing data: all those features of past landscapes that do not preserve? In some senses, the selectivity of destruction may be desirable in the sense that it focuses attention on just those aspects of the cultural landscape that are most permanent and difficult to destroy, and these are the best media to structure an institution with lasting continuity. Aspects of the environment that are ephemeral are lost to the archaeologist, but by their impermanence would have made poorer media through which to construct stable institutions. Building is of different kinds, such as large-scale constructions allowed to gradually decay versus smaller buildings added to annually as part of social renewal. The specific phasing of work and its amount can be recognized by detailed excavations (see, for example, Kolb 1991) and suggest much about the mechanism through which instituting facilities were constructed.

Second, although cultural landscapes have many aspects that are based on culturally specific meanings and interpretations, in large-scale societies the possibility for multiple interpretations of cultural symbols weakens institutions in which emerging leaders wish to impose an integrating coherence. The simplicity of the media's message is thus elemental to institutional success, communicating a common message to an ethnically and socially diverse population. The physical character of the landscape manufactured by human labor carries fairly unambiguous messages: Size tells of the labor used to build the monument, skill in execution tells of the training of that labor, walls block visual and physical access, while paths and gates make connections. The physical character of the built environment also makes it difficult to fake; although it can have different meanings to different observers, the elements of labor and control are clear and difficult to fake (DeMarrais et al. 1996). The ease by which elements of the landscape can be read by archaeologists is exactly why it is so important for institution building. The built environment tells convincing stories of human work, social relations, and hierarchy.

BACKGROUND LITERATURE

The cultural landscape has received considerable attention from both geographers and cultural anthropologists. Geographers have produced a

INSTITUTIONALIZATION OF CHIEFDOMS

long and rich literature that describes how landscapes organize space, time, meaning, and communication (see Rapoport 1994:465). Analysis has focused on the different settings of human activities and how a built environment structures the flow of action and meaning from appropriate everyday activities to socially structured ownership to cosmologies and cultural schemata.

> The marking of both boundaries and the contents of settings and domains, what are called the "mnemonic" function of cues, reduce the need for information processing; it takes the remembering away from people and puts the situations, rules and behaviors into the settings. (Rapoport 1994: 493)

Memories and associations transform spaces into places that can be marked and often reconfigured in stable forms through the built environment.

In cultural studies, time, space, and landscapes have been analyzed in terms of structure, meaning, and power. Particularly influential has been the work of Bourdieu (1977, 1979). In his analysis of a Kabyle house from North Africa, Bourdieu describes the low-wall division of the rectangular house into two sections for human and animal use. Permanent and moveable furniture that include storage jars, chests, serving tables, and looms define where family members work, meet visitors, socialize, and sleep. The relationships of activities to the consequential oppositions of meaning is suffused with contrasting lights, noises, smells, and moisture levels. The house is constructed by individuals following family goals and notions of appropriate social order. Then the house, a cultural artifact, orders the frames for daily human experience that molds the participants' ingrained dispositions. Hanks (1990) differentiates several Mayan spatial schemata that include the human body, the domestic house, agricultural fields, and ritual altars like the shaman's table. Each schemata has distinctive orientations with specific physical and cosmological references, but the different frames conceptually interpenetrate, affecting how space is used/interpreted in the other. Although emphasizing meaning, Hanks illustrates how each schemata is physically framed by objects, furniture, and structures. Bloch (1971) describes how tombs are the most solid and permanent structures that the Madagascan Merina construct. Tombs deny social fluidity; they ground personal sentiment in inalienable places, encapsulate social identity, and give an acceptable medium for the display of wealth. "By making a tomb the solidity of which gives it a durability which far exceeds everything else, the Merina affirms his allegiance to a particular social unit and an idealized way of life" (Bloch 1971:114). British postprocessual archaeologists have described, with varying success, how such socially constructed meanings and emotions were attached to landscapes of the past (for example, Thomas 1991; Bradley 1996; Thorpe 1997).

My approach to landscapes derives from a long research tradition in processual archaeology that has studied the spatial distribution of human settlements across environments (Binford 1964). Since the earliest settlement pattern research of Gordon Willey (1953), attention has focused on settlement hierarchy and their monuments. Social archaeologist Colin Renfrew (1973a), for example, used the distribution of monuments across Wessex in southern England to describe the territorial divisions of its Neolithic society into chiefdoms that persisted through time. In the Early Neolithic, several distinct clusters of burial monuments (long barrows) were each associated with special ceremonial monuments (causewayed enclosures and cursus monuments), and continuing territorial divisions became associated with later henge monuments (most famously Stonehenge) in the Late Neolithic. Many researchers estimate the labor invested in the construction of everything from European houses and long barrows (Startin 1982) to Hawaiian temples (Kolb 1994) and Mayan buildings and pyramids (Abrams 1994). Universal thermodynamic principles of the engineering costs of monumental construction make it the ideal medium for symbolic representation of power, society, and institutional durability (Trigger 1990). The scale and spatial arrangement of monuments show how labor was organized and controlled in the institutionalization of power held by leaders who would be rulers.

British social archaeologists, working in the processualist tradition, have investigated how the construction of the cultural landscape organized human groups. Burial monuments, for example, have been interpreted as defining the genealogies of corporate groups (Fleming 1973) that Renfrew has described as chiefdoms. The growth of population and/or intensification of herding apparently caused competition over the best lands on the upland chalks of Wessex. The construction of the burial and ceremonial monuments then created a cultural landscape associated with particular political groups. In some senses the monuments, their associated ceremonies, and the memories they encapsulated evidently defined the social landscape of marked territories. The soil became the home for the ancestors, and individual rights to use the land depended on the ability to trace relationships to those buried there. For the Late Bronze Age in England, Fleming (1982, 1989) describes the construction of massive "coaxial field systems" with ancient stone and earth walls that were laid out according to a unitary plan. Arbitrarily crossing the watersheds and other local topography, these constructed landscapes apparently divided space into social and political as well as economic units; they probably fixed regular patterns of use and access to land as part of a managed social map. As argued elsewhere, the evolution of chiefdoms depends on developing a managed system of land tenure (Earle 1991c). Because chiefdoms require resources to finance new ruling institutions, overarching property rights are implemented to mobilize a "surplus" of commoner produce to support the ruling segment of society.

INSTITUTIONALIZATION OF CHIEFDOMS

The emergence of chiefdoms is thus often associated with radically transformed cultural landscapes in which monuments (chiefly burials and ceremonial places) are built to define space and restrict rights of access within emerging political economies.

In a similar vein, but for simpler societies, P. J. Wilson (1988) crystallizes a new perspective on the built environment. With an implicit evolutionary framework, he looks at village formation in the Neolithic that he believes was a social transformation, the domestication of humans, and not primarily an economic revolution, the domestication of animals and plants. Human domestication involved the formulation of larger social institutions that determined much human action. New institutions of family, clan, and community were formalized in the built and stable environment of houses, paths, enclosing walls, dance grounds, and burial monuments. Who you were and how you interacted with others was not open to personal choice but was built in the stone, brick, and mud of the village walls. Excavations of village tells, for example, document that houses and monuments were repeatedly rebuilt on the footprint of earlier structures. The building of villages created the arenas that instituted tribal life by channeling daily human action and experience, and these interpersonal relationships persisted through time. The village became such a strong defining space that those living outside its walls felt unprotected by the social fabric of life, social outcasts, and vulnerable to antisocial elements like outlaws or enemies.

Fletcher (1995) sees the built environment as essential for the evolution of large-scale social groupings. Based on behavioral research, limits must exist to the scale and scope of human activities. Simply stated, complex societies that organize large numbers of people require significant institutional transformations such as the Neolithic building of village architecture and life. Barriers and conduits organize (institutionalize) human interaction in settled communities. To the degree that the evolution of human societies involves expanded scales of interaction, those interactions must be channeled. Institutionalization thus must involve physical constructions in the landscape organizing humans in such a way that radical increases in scale and complexity of social groups are practical. Social evolution must involve processes by which the natural limits on human capability and tolerance can be bridged by physically channeling interactions. Without this, settled life, villages and towns, and the development of cities would be impossible.

THREE ARCHAEOLOGICAL CASES: DENMARK, PERU, AND HAWAII

In *How Chiefs Come to Power* (Earle 1997), I compare the long-term development of chiefdoms in three independent areas of the world: Denmark, Peru, and Hawaii. The primary thrust of my argument is that social evolution in different natural and social environments follows contrasting pathways of development based on different possibilities for the central control of the economy, military, and ideology. As resource use in contrasting environments is intensified, different opportunities for control and resistance to the development of a centralized political economy emerge. Different patterns of control tie to different kinds of finance and institutional forms.

I want now to extend my original argument. Blanton et al. (1996) have suggested a useful dichotomy of social relationships: personalized networks and corporate groups (see Chapter 8, this volume). I believe that these are contrasting forms of institutionalization that develop under contrasting mechanisms of social control and finance. They should also have distinct forms of materialization.

Personalized networks involve relationships *between people.* In Renfrew's (1974) terms, these are "individualizing chiefdoms" that are both diffuse and expansive. The logic is to gain political strength by reaching out (networking in modern parlance). The political economy is based on wealth finance that involves controlling the production of prestige goods by attached specialization and/or their distribution in long-distance exchange (Friedman and Rowlands 1977). A broad (and unbounded) international class of warriors and other elites is formed through the exchange that materializes the translocal networks in what Kristiansen (n.d.; see also Helms 1979) calls the politics of distance. These networks, based on the relationships between individuals, are further materialized in burial rituals that monumentally enshrine individuals with their inalienable objects so as to solidify their social networks across time.

Corporate groups involve relationships of *people to resources* and the definition of groups with respect to those resources. Corporate groups own something—irrigation systems, fields, or the like. These groups include Renfrew's (1974) "group oriented chiefdoms," and they emphasize exclusiveness and association with place. Chiefs manage and thus control the groups' resources; staple finance (D'Altroy and Earle 1985) is based on the extraction (mobilization) of labor or produce owed to chiefs as the groups' leaders. Materialization emphasizes group ownership rights with elements in the built environment that include wall and boundary markers. Other elements include monuments, cemeteries, and ritual spaces that define patterns of inheritance and religious association.

The specific institutional character of any society will be a composite of these two instituting principles. What I want to describe is how the character of the cultural landscape of three prehistoric cases can be studied to recognize the institutional form of society and how institutions developed through time. What the archaeologist sees in the landscape is fundamentally important, because it highlights how humans constructed their environment not for the everyday, but for posterity. In simple terms, we focus on the most stable and lasting modifications to the environment that would have been basic to the construction of social institutions built to last.

The Cultural Landscape of Thy, Denmark in the Late Neolithic and Early Bronze Ages (2400–1200 BC)

During the Late Neolithic and Early Bronze Ages of Denmark, the Thy region on Jutland's northwestern edge was integrated into broad-ranging prestige good exchanges that linked chiefdoms and simpler tribal societies across much of Europe (Friedman and Rowlands 1977; Shennan 1982; Kristiansen 1984, 1998c; DeMarrais et al. 1996; Earle et al. 1998). The Late Neolithic in Thy is referred to as the Dagger Period and it is defined by beautifully flaked, lancelet-shaped flint daggers and arrow points and by bell–beaker-style ceramics. These artifacts compare closely to elements that define the bell–beaker burial assemblage elsewhere, such as in Wessex, but the daggers of Thy are not of metal and the artifacts are recovered from normal household contexts and not primarily burials. Elsewhere (DeMarrais et al. 1996), I have argued that the society in Thy was connected ideologically with a European prestige goods system; local objects (daggers, arrowheads, and beakers) were manufactured in international styles. Although a system of status rivalry must surely have characterized Thy at this time, status positions could not be consolidated because the means to materialize them were not controllable.

The construction of the cultural landscape during the Dagger Period seems to fit well with our model. In the whole of Thy only two small burial mounds can be dated to this period. The landscape was open and dotted by small hamlets or single farms located high up on hilltops, probably for the defensive advantage of visibility. Houses were often small pithouse construction, 4 m wide and 10 m long (40 m^2 roofed area); their wooden construction probably meant that use life would have been short, perhaps only a generation. The grasslands and fields were open and unmarked by constructions that would help institutionalize/stabilize interpersonal relationships. We can imagine an open social field of status rivalry, warfare, and defense.

During the subsequent Early Bronze Age, the nature of society in Thy was transformed. The artifact assemblage changed quite dramatically. The elaborate flint work in the daggers and arrow points ended and knapping was geared to a range of expedient flakes, scrapers (hide working), and sickles (cereal harvesting) (Aperlo 1994). Ceramics became largely undecorated. Housing styles changed; larger post-and-beam, three-aisle houses replaced the small pithouses. The normal house was perhaps 6 m wide and 15 m long (90 m²), and chiefly houses range up to 34 m long and 8 m wide (272 m²) (Fig. 1). Contrasting house sizes suggest a hierarchical society, apparently based on the production and export of animal products and amber. The 34-m long chieftain's house in Sønderhå parish sits on a rise in the landscape, below a cluster of contemporaneous barrows. Twelve pairs of massive wooden posts held the frame, and the walls were constructed of wooden planks. The plentiful use of wood was lavish in the extreme; at this time in Thy, the landscape was largely grasslands, depleted of its former forests; the wood for such a house would have had to come from specially owned (and sacred?) stands, from limited driftwood collected along the North Sea, and/or from wood cut from Norwegian or other distant forests.

Bronze was imported and manufactured locally for warrior and chiefly swords (Kristiansen 1982) and some jewelry. These were customary objects of prestige, probably manufactured by attached specialists, given within

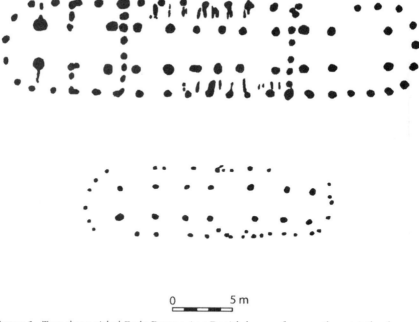

Figure 1. Two three-aisled Early Bronze Age Danish houses from northwest Jutland: upper is a chieftains house at the Legård site in Sønderhå (Kristiansen 1998c:Fig. 4a); lower is a more typical house from the Glattrup site in Skive (Bertelsen 1994:180). Both houses have similar plans, differing primarily in size (34 m vs. 24 m in length).

exchange networks linking elites together, and closely associated with a person's identity and buried with him at death. The prestige goods materialized a person's status as set by his relationships in political networks that stretched across northern Europe.

During the Early Bronze Age (periods II and III), the construction of a cultural landscape in Thy shows some continuity with the later Neolithic, but with dramatic changes tied to the institutionalization of "individualizing" chiefdoms. Continuity is documented in the settlement of single farms in an open landscape of grasslands for grazing and of some fields. The wood-framed houses were short lived and their situations shifted around within a locality (Mikkelsen 1996). Field and territorial borders were unmarked in any recognizable way.

The transformation of the cultural landscape focused on the construction of probably 2000 burial mounds within a few hundred years.

> Certain historical epochs have put a special mark on the landscape that has lasted until today, and make us wonder what made people demonstrate such energy.... In Thy..., the landscape is even today a barrow landscape—wherever you turn, the eye meets one or several barrows on the horizon. People during this period were certainly self-conscious and proud, but how are we to understand the rationality behind... such massive ritual investments? (Kristiansen 1998c: 281).

At the center of the mound a stone cyst was placed into which the primary inhumation burial or cremation burial and prestige goods were laid (such as the warrior or chiefly swords). From the surrounding pasture land, turfs were cut and mounded over the cyst, and the barrow was edged by a low stone curbing. Standard diameter of a mound was about 15 m and estimated height 2–3 m. Within most parishes, significantly larger mounds (over 4 m high) existed. Estimated labor in the construction of an individual mound were perhaps 1000 man days, significantly more for the larger mounds.

The construction of the mounds has permanently altered the landscape of Thy. Across perhaps 10 generations, several thousand mounds were built to mark where individuals were buried. These mounds often were located on ridge crests, creating a distinctive profile against the sky that is still visible for several kilometers (Fig. 2). Based on the circumstance that mounds were clustered in groups, were often rebuilt, and have later burials added into them, I believe that specific locations were associated with corporate kin groups, perhaps lineages. Such lineages could have been associated with chiefly lines; certainly the burials with their swords and other bronze finery indicate individuals of high status. Their mounds express spatial relationships of ancestors to the land, just the type of permanence described by Bloch (1971). I believe that the Bronze Age landscape of Thy typifies a

Figure 2. Early Bronze Age barrow group in Kregme, Denmark (1887 watercolor). Each mound marks the grave of at least one chieftain or warrior (Courtesy Danish National Museum).

chiefly institutional form in which individuals are distinguished by monumental burials so that their personal histories are remembered.

A Bronze Age landscape of individual barrows has also been described for southern Britain. There, a dramatic transformation in the landscape was associated with the transition from Bronze Age to Iron Age societies that may signal a fundamental political change in institutional formations and finance (Earle 1991c). In the individualizing Bronze Age chiefdoms, the burial monuments were prominent at the same time that there were no marked settlements. Then, in the Iron Age, a settlement hierarchy developed with prominent hill forts encircled by substantial walls; however, no monumental burials or other monuments existed. The same scale of labor as had constructed the barrows and henges was then dedicated to the construction of settlement fortifications. In the Iron Age, the landscape of southern England also was transformed by grids of agricultural fields. These may well have been assigned to individual farming households required to supply surpluses as rent to an overlord; the surpluses were then apparently held in storage facilities located centrally within the main hill forts (Gent 1983). These hill-fort chiefdoms, based on staple finance, find parallels around the world as in highland Peru during the Late Intermediate Period.

The Cultural Landscape of the Mantaro Valley, Peru in the Wanka II Period (AD 1300–1460)

During the period just prior to Inka imperial conquest (Wanka II period, AD 1300–1460), the Mantaro Valley in Peru was organized into multiple competing chiefdoms. The Upper Mantaro and Yanamarca Valleys are intermontane. The valley floors below 3400 m are farmed for maize and now wheat, the immediately surrounding hills are covered with extensive fields especially for potatoes and quinoa, and above 4000 m the puna grasslands are grazed by Andean camelids (llama and alpaca) and now sheep (see Hastorf 1993).

The indigenous local society, the Wanka, have a long history dating from the Middle Horizon (about 800), and their material culture included distinctive bi- and tricolor storage pots and plainware, micaeous-tempered cooking vessels (Costin 1986; Hagstrum 1986). Lithics included distinctive hoes manufactured from large flakes struck from stream cobbles and chert blades from local sources (Russell 1988). Silver and copper from regional mines were manufactured locally into a variety of decorative (pins and disks) and utilitarian objects (especially needles). Most distinctive of Wanka culture were large, fortified settlements placed high on ridges and hills above the valley floor.

I have called the Wanka a hill-fort chiefdom, quite similar to such societies in southern England and elsewhere (Earle 1997). Characteristic of these chiefdoms is a staple finance economy in which land is held by local chiefdoms. Several Wanka hill-forts, of different sizes, composed a compact polity with centers sometimes only 10–20 km apart. Upward of 10,000 people can be concentrated in one settlement. Each hill-fort massed people in a highly defensive and fortified position that would have been all but impossible to defeat by frontal attack. In the resulting political stalemate, polities of 10–20,000 were holed up in their small clusters of hilltop redoubts.

In terms of institutional materialization, the pattern fits well with a corporate principle of organization. Status differentiation of individuals is marked only in subtle ways. Prestige goods, such as the metal objects, were more concentrated in elite households, which themselves were marked by better construction and central location, but the differences were in degree and not kind. Burials were not individually distinguished or monumental; the dead were interred either in the subfloor of houses or in closed burial caves, and no special differentiation of status was noted in the burial goods (Owen and Norconk 1987).

The distinctive and easily recognized form of materialization was in the layout and building of the settlements (Fig. 3) (DeMarrais n.d.). The primary social units of the household and the settlement were clearly structured by the architecture. The household was defined by one to seven circular stone houses with doors that opened onto a patio work space. The patio was further demarcated by stone walls that permitted access through narrow gaps between buildings or through the walls. View into or movement through the private spaces of each household were obstructed by these stone walls.

Houses were then jumbled together within the settlement. An irregular

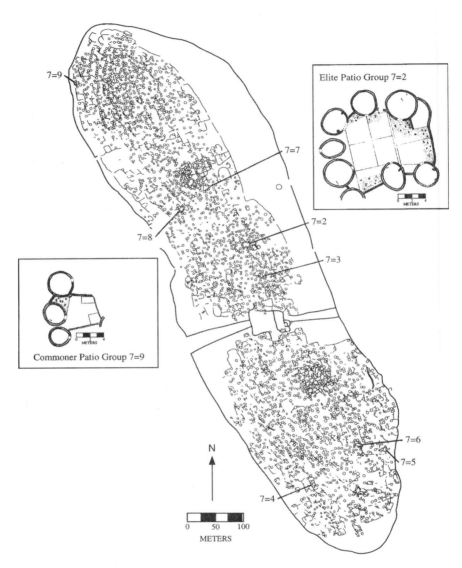

Figure 3. Late Intermediate period hilltop settlement of Tunanmarca in the Yanamarca Valley, Peru. Note the surrounding fortification walls, the central plaza, and the corridor that divides the settlement in two. Two patio groups are enlarged: 7=2, an elite household with six houses; 7=9, a commoner household with one house.

INSTITUTIONALIZATION OF CHIEFDOMS

network of paths threaded between the houses, permitting movement through the settlement. No organization was evident from the paths, along which visibility was limited. They were not organized according to any recognizable grid, symmetrical, or radial pattern. Apparently settlements were not internally differentiated, except at major centers where a rough bilateral symmetry existed. At Tunanmarca, for example, habitation areas of approximately equal sizes flanked the small plaza on the ridge's highest point; this arrangement may have represented the Andean moiety system. Community ceremonies probably took place at the central plaza; however, the size of the ceremonial ground was insufficient to hold more than a fraction of the settlement's total residential population. No monuments or public buildings characterized these large settlements.

The primary architectural definition of the settlement was the surrounding defensive walls that were probably built for protection by community labor. A discussion exists as to whether defensive walls of such settlements served to exclude an enemy or to include and define a social group as insiders, if you will (see Adams 1966; Earle 1997). Although the fortification was most likely built when real threats existed, the fortification certainly defined the social group. Warfare acts to bring people together for mutual defense reinforced by the common settlement's walls that limit movement and define the settlement unit.

Wanka communities often were associated with fairly intensive agricultural systems (Hastorf and Earle 1985). These included irrigation systems that carried water from springs and streams, hillside terraces, drained field on the valley floor, and ridged fields on the edges of the puna. Agricultural intensification (marked by the channels, ridges, and retaining walls) created a physical landscape that often is associated with intensification and explicit systems of land use (Stone 1994). Such ownership is basic to staple finance systems, in which access to land is given to commoner households in return for their corvée labor or for produce from their fields. The agricultural facilities were probably constructed by social labor organized by the local leader, who then assigned rights of use to the plots marked by the walls and terraces. In the Andes today, the irrigation ditches are cleaned annually by the community in special ceremonies that emphasize the vital supernatural forces of the water and fertility (see Mitchell 1991). Membership in a community and rights to use community lands require participation in such work projects. The importance of social labor for constructing agricultural landscapes that support this kind of staple finance is further dramatically illustrated by the historic chiefdoms of the Hawaiian islands.

The Hawaiian Cultural Landscape during the Formation of Regional Chiefdoms (AD 1200–1800)

By AD 400, Polynesians had discovered and colonized the Hawaiian islands, isolated in the deep Pacific far from other island groups. For the next 1000 years, population grew, and largely in isolation Hawaiians formed several large and complex island chiefdoms. Paramount chiefs and a genealogically designated chiefly hierarchy ruled and fought over the land and its people. These chiefs were gods on earth, separated socially and religiously from the commoners. They lived in special houses and conducted elaborate ceremonial cycles at the temple grounds where their gods resided (Valeri 1985). The political economy was based on staple finance, by which the chiefs (owners of all lands) mobilized foods that supported the personnel of the chiefdom including the chiefs, land managers, warriors, priests, and attached craft producers (Earle 1978). Commoners lived on the land, receiving specific farm plots and a house lot in return for the contribution of labor and goods to the chiefly owners of the communities. Hawaiian subsistence came from intensive taro and sweet potato farming, from fishpond and sea fishing, from the raising of pigs, and from foraging.

In terms of institutional materialization, Hawaiian chiefdoms represented composite networking and corporate principles of organization that probably characterize most complex chiefdoms and states. Social status within the chiefly hierarchy was signaled by distinctive dress, most dramatically the feather cloaks and helmets that identified chiefs with their high gods (Cummins 1984; Earle 1990). These objects of personal adornment were manufactured by attached specialists and given politically to establish relationships in the governing hierarchy. Such materialization was flexible and impermanent; the defeated chief was stripped of his adornments. Monuments did not distinguish the burials of individual rulers, whose bones were either hidden away or kept in chiefly *heiaus* not specifically associated with the chief.

The primary form of institutional materialization was the cultural landscape. As I have argued elsewhere (Earle 1997), the intensive field systems created a carefully measured and marked landscape. Irrigation systems tied together strings of taro, pond, fields, fishponds, and adjacent dry farming for sweet potato, coconuts, sugarcane, and other crops (Fig. 4). Fields were defined by permanent walls of stone and earth that were constructed socially by the community and organized by the chief's managers (Earle 1978, 1998). Construction was a social event when all turned out to work together; the built facilities for farming were both clearly visible and permanent. The landscape was neatly divided and rights to it were assigned by

INSTITUTIONALIZATION OF CHIEFDOMS

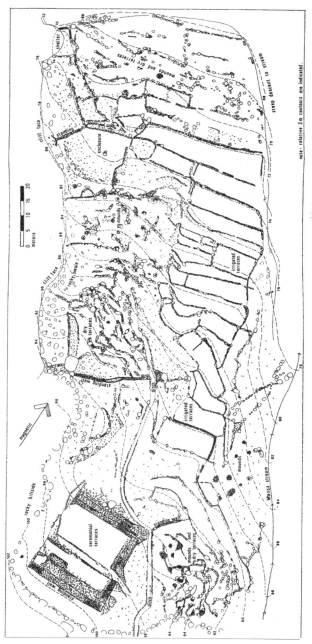

Figure 4. Site K-D5-8, Ha'ena ahupua'a, Kaua'i, Hawai'i. An early historic irrigation complex with agricultural terraces, a small heiau, habitation, and dryland farming area. The system's ditch was fed from the small Manoa stream (Earle 1978:Fig. 6.6).

the chief's managers. In exchange for labor and goods given to the chiefs, commoners received use of land plots for a house lot and farming. These plots were not abstract rights; they were physical segments of the fields and land divided by walls and ditches. The intensive farming based on large facilities created a permanent physical landscape that encapsulated the rights and responsibilities in the system of ownership on which the Hawaiian political economy was grounded.

The Hawaiian social system itself was materialized in the constructed landscape. The family held a private house lot, set off by an enclosing wall. Within the *ahupua'a* communities, local segments were defined by walls, trails, and shrines that were the physical map of and channels for social interactions. Daily activities took place on the constructed landscape such that the routines of life became associated with the physical constructions fairly permanent across the generations. The community itself was often defined by a natural unit (the valley) with the dividing ridges marking the boundaries. Along these ridges, each distinctive rock and geological feature had a name and was associated with mythical and historical narratives. The geography was further marked by walls and rock art sites. At the boundary between the community was a shrine where during the annual *makahiki* ceremony the community's members were responsible to deliver food and goods to the gods represented on earth by the high chiefs.

The regional organization of the Hawaiian chiefdoms was materialized and ceremonially linked to its ruling ideology (Earle 1997). This included a paved coastal trail and the ritual landscape of chiefly and community *heiaus*. The materialization represented hierarchy and central integration through shrines (identified with the levels of local segment, community, and region) that were positioned regularly across the landscape. Framed by the *heiaus'* architecture of enclosing wall and platforms, and visually marked by towers and buildings, the ceremonies incorporated the gods present on the *heiau* in the form of stone uprights and carved or woven images (Fig. 5). Two major types of *heiaus* and related ceremonies involved chiefs in rituals of war (with human sacrifice) and of fertility. The ceremonies related to war, agriculture, politics, and song created overlapping but not entirely reinforcing institutions with associated nodes of power.

Kolb (1991, 1994, 1997) has systematically documented the long-term construction of the Hawaiian cultural landscape as part of the development of regional and island chiefdoms. By combining a description of the different elements of the constructed landscape (shrines, walls, paths, agricultural fields, etc.) and the dating of building phases, he documents how the social institutions of the chiefdoms were incrementally built. Much of the building activities were concentrated in a short time span during the forma-

INSTITUTIONALIZATION OF CHIEFDOMS

Figure 5. A religious compound (heiau) at the Waimea ahupua'a, Kaua'i, Hawaii. The engraving is after a sketch made during the visit of Captain James Cook, European "discover" of the Hawaiian islands (Cook 1784).

tion of the regional chiefdoms. The chiefdoms, created by conquest, were institutionalized by the construction of facilities and monuments that then were the stages for major ceremonies. Through time, different elements of the community were literally built up to determine, emphasize, and solidify social relationships, religious dependence, rights of resource use, and political subordination.

The potential exists to describe step by step the building of prehistoric landscapes. All environments have meaning to the people who inhabit them, but that meaning can be altered strategically through the built environment. Some media are accessible to all—rock art or stone cairns. They can be easily made and remade. Other media require organized labor to fell large trees, move heavy stones, and craft special embellishments. Standard archaeological investigations that require survey, stratigraphic excavation, and routine dating of construction phases can define the steps taken by communities and chiefly institutions to build the facilities that support the stable social relationships on which complex societies are made. Using labor as a universal currency (Trigger 1990), we can unfold the specific phases in which the built environment was raised and renewed within the ideological systems of particular times and places.

CONCLUSIONS

I have described a simple contrast between (1) chiefdoms organized by principles of networks that emphasize individuals and interpersonal relations and (2) chiefdoms organized by corporate principles that emphasize the relationships of people to land or other things. The Bronze Age chiefdoms of Denmark relied heavily on networking and their finance was based on wealth finance. In contrast, the Wanka chiefdoms of Peru relied on corporate principles and their finance was based on staple finance. The complex chiefdoms of Hawaii combined both organizing principles and relied on staple finance to fund the manufacture and distribution of wealth.

The contrasting principles of organization and their linked forms of finance were institutionalized by distinctive forms of chiefdoms that were materialized in different ways. The Danish chiefdoms built personal burial monuments prominently placed in the landscape, and individuals and their relationships were materialized in distinctive prestige objects. Totally differently, the Wanka chiefdom marked the landscape with the facilities of intensive agriculture, and the communities were built in stone with walls that marked the community off and identified clearly the individual households of the community. But no individual burial monuments and very little wealth characterized the society. The complex Hawaiian chiefdoms, like the Wanka, emphasized the social construction of a built environment, but the extent and complication of relationships were more elaborated. Each wall, trail, terrace, and canal was built by social labor and concretized social relationships across time. But in addition a more dynamic and flexible system of hierarchical network was built and displayed through production and gifting of wealth objects.

Chapter 7

Cosmology and the Institutionalization of Hierarchy in the Maya Region

PATRICIA A. MCANANY

Shamanism, some would argue, is the key to understanding classic Maya rulers (Freidel 1992; Freidel et al. 1993). But what does kingship have to do with priests, healers, witches, and seers who stand with one foot in society and the other in an ethereal liminality? Animal spirits, ancestors, and supernaturals would seem strange bedfellows for the worldly king who wielded power, organized military forays, and extracted labor services from his populace. Years ago, Eliade (1970:333–336) wrote of shamanism as the ecstasy of religious expression, hinting at the deep, Asian roots of shamanism in the Americas. Likewise, ethnographers, have enjoyed a long tradition of anthropological investigation of shamanism among Native Americans (see Furst 1974, among others). But in neither case was shamanism linked with power of an earthly kind or viewed as a pathway to rulership. More recently, however, scholars have proposed a wedding of shamanic power

PATRICIA A. MCANANY • Department of Anthropology, Boston University, Boston, Massachusetts 02215.

From Leaders to Rulers, edited by Jonathan Haas. Kluwer Academic/Plenum Publishers, New York, 2001.

with political authority to produce "shaman kings," or as Chang (1983:112) has written in reference to state formation in China, "shamanistic politics." This approach, which models societal cosmology as having been deftly shaped into political ideology, suggests that the transition from leaders to rulers is as much about how we think as it is about what valuables we possess, with whom we ally, or who we can best militarily.

Such an ideational approach to political transformation can be at odds with materialist-based models of political change or can work in a complementary fashion to provide an explanatory narrative that fills some of the loopholes of a strictly materialist approach. Here, I attempt to work in the latter mode, critically evaluating the notion of "shamanistic politics," while at the same time integrating cosmology with political authority. In order to do so, I first examine two concepts central to rulership: power and time. Then, several factors argued to have been seminal to the rise of political authority are scrutinized. As my case study, I examine the rise of kingship in the Maya lowlands.

POWER, POLITICS, AND TIME

Who is a ruler but a person with institutionalized political authority? While the authority of any one individual ruler may be contested, the position, the office held by a ruler, generally is not. Being human, we cannot help but ask how this situation germinated from the "seedbed" of Formative or Neolithic village life. Older ideas of the egalitarianism of early villages have gone by the wayside and we can assert with reasonable veracity that kingship was born in the hearth of village life. Many scholars (particularly those bred on a secular style US democracy) have asked why village societies allowed the loss of "voice" inherent in accepting a ruler or overlord and many theories stressing either the putative coercive or cooperative aspects of the process of political centralization have been proposed (e.g., Blanton 1983; Carneiro 1970; among others). Adopting the paradigm of structural Marxism, one can conjecture that the loss of village autonomy was the result of a short-term solution that held unforeseen long-term consequences (à la Pauketat and Emerson 1999). Alternately, the process may have been somewhat irrefutable and incontestable, on principle if not in reality.

The "irrefutable" aspect to which I refer is a kind of naturalization (or more accurately, culturalization) of power—power as defined by the concentration of potent cosmological forces. This type of power is distinctly non-Western in conception and praxis. A few years ago, I suggested (McAnany 1993) that the sources and manifestations of power, status, and wealth in Classic Maya society were diverse and not necessarily congruent. For instance, a successful commoner merchant may possess more wealth than the youngest sibling in a minor élite family. While total congruence should never be assumed, nevertheless, I now see power in Classic Maya society in more monolithic terms that are analogous to Anderson's (1972) discussion of power in Javanese culture. Unlike secular notions of power in Western society, power in the Javanese tradition is concrete (an existential reality), a homogeneous entity (despite the identity of the ruler who is wielding it), a fixed and constant amount (only its distribution varies), and most importantly, power exists without questions of legitimacy (Anderson 1972:7–8). This conception of power is based on the Javanese notion that power or radiance (*tédja*) flows from a connection with cosmological forces of the primordium, which itself is a concrete and constant quantity that exists without reference to morality or legitimacy. Possibly related conceptually is the Classic Maya term *ip* which is often translated as vital force or power (Houston and Stuart 2000). As we shall see, the manner in which Classic Maya rulers linked themselves to supernatural forces suggests that political authority—manifest in the institution of kingship or *ahau* and commencing no later than the third century AD—was fabricated, at least in part, on a "culturalization" of the forces of the cosmos (Freidel et al. 1993:58).

Knowledge of the cosmos, though, need not be linked with political power, and indeed is not in many societies. In an effort to understand the transformation of cosmological precepts before and after the time of rulers, Helms (1999) has proposed a heuristic model of two temporally sequential cosmological axes. The older, spatial axis is manifest in rituals enacted to contact the world of animal spirits, supernaturals, or generalized ancestors. It is the realm of shamans living among mobile peoples, such as those studied by Eliade (1970) and others. The second cosmological axis originates with village life when a new emphasis on tenurial issues begets "house"-focused societies (à la Lévi-Strauss, 1982:174–187). This second, temporal axis situates a fixed house and its surrounding landscape as a point of origin (and a pathway) to a past of named ancestors, supernaturals, and ultimately to the great creation events. Through the temporal axis, ancestors come full circle and merge with supernaturals and spirits arrayed along the spatial axis. Helms (1999) contends that the ontogeny of the temporal axis is at the root of wealth accumulations and the institution of rulership. Part of the key to the asymmetrical significance of this second axis lies in the fact that it allows certain families greater "access to first principle origins"—the armature of house-based societal structure. To Helms, then, it is the politicization of *time* that facilitates the centralization of power. This model is well-matched to lowland Maya society in which, as we shall see below, great cosmological events folded within cycles of time were of central concern to emergent Maya rulers.

Grappling with these central issues of *time* and *power* in ancient China, Chang (1983:107) concludes that there are seven factors critical to the rise of political authority. His analysis, based on both ancient Chinese texts and archaeological remains, yields a suite of causal and preexisting factors that include the following: (1) hierarchical organization of kin groups with individual status roles, (2) interactive regional polities that mutually reinforce each other, (3) a military tradition, (4) narratives stressing actual or mythical meritorious deeds of past leaders, (5) presence of writing to validate one's position in a kinship system and one's access to ancestral wisdom (the latter often cited as the key to governance), (6) exclusive access to ancestors and supernaturals through shamanistic ritual, and (7) wealth and its aura. Chang does not propose this multivariate group as a universal explanation for the emergence of rulers. China's deep tradition of literacy with 20 centuries of written records as well as a distinctive iconographic tradition no doubt are the result of a distinctive cultural fabric not necessarily observable elsewhere. On the other hand, Maya society, with 13 centuries of pre-Hispanic written texts and an equally rich iconographic tradition, may have etched a similar path through the time–space continuum. In the spirit of inquiry into the ontogeny of rulership, these seven factors, so compelling to Chang, are examined for their resonance in the Maya region. Also scrutinized is whether this set represents causal factors or merely symptoms of a process already *fait accompli*. But first, space–time systematics and relevant site names and terminology for the Maya lowlands are presented.

FORMATIVE TO CLASSIC PERIOD IN THE MAYA LOWLANDS

By convention, the transition between the Formative and Classic periods is dated to AD 250. This date is linked to one of the oldest long-count dates in the Maya lowlands, specifically stela 29 from the site of Tikal (Jones and Satterthwaite, 1982). On the front of this stela, Scroll-Ahau-Jaguar wears full ritual regalia. The reverse side contains a single column long-count date of 8 baktun, 12 katun, 14 tun, 8 uinal, and 15 kin, or AD 292 (Fig. 1). Thought to represent a royal accession event at Tikal, stela 29 does not inform us as to how institutionalized rulership came about; rather, it signals the end of that transition. The tradition of erecting stelae bearing long-count dates and hieroglyphic texts that generally were associated with rulership continues through the Early Classic (AD 250–550) and the Late to Terminal Classic (AD 550–900/1000) periods. At the end of the Classic period, there is a fundamental transformation in the nature of statecraft in the Maya lowlands. The old dynasties appear to collapse and polity size shrinks, in some cases, to the scale of community (see Marcus 1998, for more on cycles of state development).

INSTITUTIONALIZATION OF HIERARCHY IN THE MAYA REGION 129

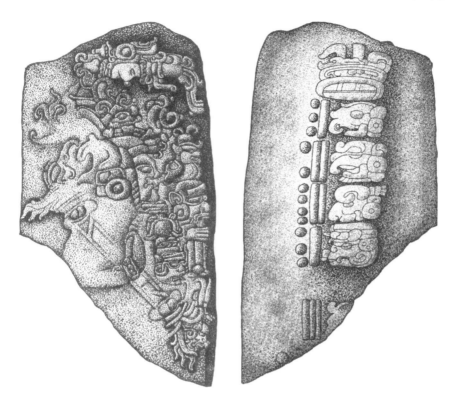

Figure 1. Stela 29 from Tikal showing Scroll-Ahau-Jaguar on front and the long-count date that marks the beginning of the Classic period on back (drawing by J. A. Labadie after Jones and Satterthwaite 1982:Fig. 49).

To understand how the kings of the Early Classic gained their mandate to rule, we must look back to the Formative or Preclassic period, which in the Maya lowlands generally is divided into a Middle Formative (1000 BC–400 BC) and a Late Formative (400 BC–AD 100). The remaining 150 years (AD 100–250) are either separated into a Terminal Formative period or called the Protoclassic period. In contrast to the vibrant Early Formative settlement on the nearby Gulf and Pacific Coasts (Blake et al. 1995; Clark 1997; Rosenswig 1998; Stark and Arnold 1997), very limited evidence of settlement in the Maya lowlands during the Early Formative period (2500 BC–000 BC) has been forthcoming. Around 1000 BC, expansion into the Maya lowlands by village farmers seems to have been swift and decisive and population levels built quickly after initial settlement of prime locales (Andrews 1990; Hammond 1991; among others). The material possessions of these early presumably

Mayan-speaking people included well-built pottery, a tradition of papermaking, maize- and orchard-based agriculture (including *cacao*), an elaborated mortuary tradition, and some measure of social ranking.

Prime locales of settlement included the river valleys of Belize, the Department of El Petén, Guatemala, and southern Campeche. Moreover, the span of time between earliest settlement and earliest construction of monumental architecture generally is less than 500 years. In Belize, Formative villages at which modest pyramids or large platforms were constructed at the end of the Formative period include Cuello, Cerros, Nohmul, and possibly K'axob and Colha (Fig. 2). Construction of massive pyramids occurs during the Late Formative period at key locales such as Lamanai (located on the New River Lagoon of Belize) and at the northern Petén sites of Nakbe, Tintal, Wakna, and El Mirador (Hansen 1998) (Fig. 2). The latter sites feature a distinctive triadic arrangement of pyramids most of which are greater than 30 m in height. During the Late Formative period, triadic groups also were constructed at burgeoning settlements both to the north at Calakmul and to the south at Tikal, Uaxactun, and Sacnab (Hansen 1998). As Laporte and Fialko (1990:64) have reported, another configuration of monumental architecture sometimes referred to as the "Commemorative Astronomical Complex" or "E-Group Complex" was constructed coevally at Tikal and Uaxactun and possibly Yaxha and Balakbal, with earlier examples to the north at Nakbe and Wakna (Hansen, 1998) (Fig. 2). Although pyramid construction continues through the Classic period, some key differences, discussed below, separate Formative pyramids from those of the Classic period.

Although monumental architecture provides a clear indication of an ability to amass a large labor force, it does not necessarily signal the presence of a ruler. Palaces as well as elite iconography and hieroglyphic texts are more direct indicators of kingly prerogative. As Sharer (1992:134) has noted, few of these indicators of "elite subculture" existed during the Formative period. The critical observations that can yield insight into the ongoing process by which hierarchy was institutionalized in the Maya lowlands therefore are to be found in the mortuary practices and architectural differentiation of the Formative period, as observable at both large and small centers of settlement.

KIN HIERARCHIES AND INDIVIDUAL STATUS ROLES

This precondition of political authority is to be sought out in the early villages, the crucible of kingship. For instance, Chang (1983:122) notes the structural pattern of some Yang-Shao villages in which houses are grouped around a central plaza space and differential mortuary customs mark certain individuals as those of exceptional status.

INSTITUTIONALIZATION OF HIERARCHY IN THE MAYA REGION 131

Figure 2. Location of key Maya sites mentioned in text.

Characterizing Formative-period village structure in the Maya lowlands is a difficult undertaking, since pronounced continuity in settlement locale from the Formative through the Classic period means that Formative settlements often are overlain by significant Classic period "overburden." Important exceptions to this rule exist, however. Northern Petén sites such as El

Mirador and Nakbe experienced only limited, Classic-period growth following exuberant, Formative period construction activity. Furthermore, many northern Belizean sites, such as K'axob (Fig. 2), were not abandoned during the Classic period, but programs of monumental and residential construction that were undertaken were relatively modest in scale (e.g., pyramids less then 20 m in height), permitting access via excavation to underlying Formative period deposits.

At K'axob, examination of Formative village structure through a series of large, horizontal excavations reveals the progressive expansion of the village from a core area initially settled around 800 BC to a pattern of a central, alpha residence surrounding by smaller, more modest satellite residences (McAnany 1995; McAnany and López Varela 1999). Stratified floors of a 64-m^2, apsidal-shaped, Middle Formative residential unit occur only within one excavation locale (Operation I), which is located underneath the southern pyramid plaza of K'axob. Around 200 BC, the repetitive activity of reflooring with minor expansions at this locale was abandoned in favor of the construction of a meter-high rectangular platform. At this same time, new construction of modest apsidal structures commenced at locales 100 to 200 m distant from the alpha structure within Operation I. These fine-grained data on Formative residential patterns are a rare occurrence in the Maya lowlands; such information allows the recognition of intrasettlement hierarchies expressed in residential differentiation. Although K'axob ranks as a small village compared with coeval centers of the Petén, which boasted large triadic complexes of pyramids, nevertheless the processes observable at K'axob are highly relevant to those observable earlier elsewhere. If we assume furthermore that the satellite residences were occupied by members of the same kin group as were living in the core, it appears that a kin hierarchy developed at this time. Within the core itself, which is later buried beneath a Classic-period pyramid complex, there is pronounced differentiation in individual status, expressed in mortuary practices, beginning about 600 BC, the earliest phase of occupation.

The Formative villages of the Maya lowlands then were not egalitarian villages. At Middle-Formative K'axob, evidence suggests that a sense of "house" as the physical embodiment of a corporate group with emphasis on descent from an apical ancestor is strongly entrenched in the first settlers. One of the earliest Middle Formative burials, interred underneath the initial house floor of the core area, was an single, adult male buried with 2000 shell beads and two imported bichrome pottery vessels (McAnany et al. 1999). During this time period, no other burial of equal elaboration has been found, which suggests that at the time of the mortuary ritual this individual was perceived as a founding ancestor. This type of mortuary ritual, one that stresses individuals and creates ancestors, became more elaborate through the Formative period; in fact, deceased village leaders who were destined to become ancestors may have begun to merge with supernaturals. This notion is supported by directional change in death ritual; around 200 BC there was a trend away from extended burials (which were probably loosely shrouded) to tightly wrapped seated and flexed burials that could be carried in ritual processions and displayed for a period of time before final interment. At the end of the Formative period, emphasis on ancestors culminated in the collection and reinterment of select ancestral bones at focal locales prior to building a nonresidential, monumental structure. This type of secondary interment is common in northern Belize, occurring at Altun Ha (Pendergast 1982), Colha (Sullivan 1991; Wright 1991), Cuello (Robin 1989), and K'axob. In the past, this pattern often has been interpreted as indicative of human sacrificial burials, but on closer inspection of skeletal evidence it has become apparent that many of the bones are badly weathered and the skewed elemental representation indicative of secondary interment rather than human sacrifice. These deposits suggest that bones of the ancestors "paved the path" to the institutionalization of religious power represented by pyramid construction. Although it is difficult to determine the extent to which these ancestors were perceived as divine or at least in the company of supernaturals, this gathering of ancestors, often around one or more primary interments, is a practice that was repeated often through the Classic period, ostensibly to link primary interments with venerated ancestors. This ritual practice was not unique to the Formative Maya; Brown (1997:479) discusses the tradition of reinterring the remains of elite ancestors in focal shrines at Mississippian sites in North America. In summary, the critical role of ancestors as links to the great primordium (the ultimate source of power) as well as conduits through which wealth and privilege were inherited is well established by the Late Formative period.

During the later Classic period, individuals of great status were memorialized through monumental architecture. In effect, pyramids became personal funerary shrines for rulers and their families. Unlike the funerary shrines of Old Kingdom Egyptian pharaohs, however, Maya pyramidal shrines were built within the core settlement area and often were placed on an axial alignment of historical significance. Classic period scribes used the term *wits* (meaning mountain) to describe these pyramids, some of which were named in hieroglyphic texts after ancestral kings who may have been buried within (Stuart and Houston 1994). A case in point is provided by the Temple of the Inscriptions at the site of Palenque. While restoring the temple and its underlying pyramid, Alberto Ruz Lhuillier discovered the royal tomb of Pakal the Great who had ruled Palenque from AD 615 to 683. The Temple of the Inscriptions, now thought to have been completed by Pakal's son, was so called because it contains one of the longest intact hieroglyphic texts (617 glyphs) recorded in the Maya lowlands. Carved into three panels on the inner walls of the temple, the inscription chronicles the dynastic history of the rulers of Palenque up to the time of Pakal's son, K'inich Kan B'alam II (Robertson 1983). The west panel is particularly revealing as it ends with an identification of the temple–pyramid as the *muknal* or burial place of K'inich Janahb' Pakal of Palenque (McAnany 1998) (Fig. 3). This example explains why the Classic Maya lowlands often are distinguished as a locale of the "identified ruler" (Grove and Gillespie 1992:35–36) in contradistinction to other parts of Mesoamerica such as the highland state of Teotihuacan where images and tombs of rulers are rare to nonexistent (Cowgill 1997; Chapter 8, this volume). While Feinman (Chapter 8, this volume) attributes this distinction to the stronger corporate dimension of the political fabric of Teotihuacan, Cowgill (1997:154) suggests that it indicates less emphasis on inheritance and validation of rulership through pedigree. In reference to the latter observation, it is highly pertinent that a strong tradition of kin hierarchy, individual status marking, and ancestor veneration can be traced to the Late Formative period in the Maya area.

INTERACTIVE POLITIES THAT MUTUALLY REINFORCE EACH OTHER

Chang's second factor hints at a preexisting leadership structure that is more fluid and open than that of rulership yet territorially constituted so as to be recognizable as a polity. This notion, later codified by Renfrew and

Figure 3. Final portion of the West Panel of the Temple of the Inscriptions at Palenque which identifies the structure as the burial place of K'inich Janahb' Pakal, divine lord of Palenque (drawing by J. A. Labadie after Robertson 1983:Fig. 97).

INSTITUTIONALIZATION OF HIERARCHY IN THE MAYA REGION 135

Cherry (1986) as "peer–polity interaction," emphasizes the transformative potential of situations in which there are multiple, interactive agents in a manner that is somewhat similar to complexity theory. Setting aside the problem of whether or not polities can exist in the absence of rulers, nevertheless we can examine the Formative Maya data for evidence of multiple and interactive seats of power or at least material patterns delineating and differentiating centers. In fact, evidence of several different sorts exists. One of the roles of monumental architecture, beyond the definition of sacred space, is to mark a seat of power (even if that power is the ability to summon ancestors and supernaturals from the top of a newly built pyramid). Such concentrations of power in fact did gel by 400 BC, as indicated by vigorous pyramid building in the northern Petén at the sites of Nakbe and the awe-inspiring El Mirador (Hansen 1991; Sharer 1992), with its massive triadic pyramid complexes, the largest of which (the Danta complex) attained a height of nearly 70 m. Excavations in the core area of El Mirador (which covers roughly 2 km along an east–west axis) affirm Formative period construction and abandonment (Matheny 1986; among others), which leads to the inescapable conclusion that the largest construction ever conceived and executed in the Maya region occurred during the Late Formative period. Furthermore, El Mirador is linked with Nakbe and other Late Formative sites [possibly even Calakmul (Folan et al. 1995a)] by way of a raised causeway or *sacbeob*, demonstrating that these sites were highly interactive. Whether these causeways provided a pathway for ritual processions or economic exchange or both is unknown; however, the causeways do seem to be graphic evidence of an emergent "network strategy" (Blanton et al. 1996:4) linking well-defined seats of power.

During the ensuing Classic period, the pattern of interactive polities that mutually reinforce each other can be observed in the appearance of compound hieroglyphs that identify seats of power or royal titles. Recognized originally by Berlin (1958), such glyphs are generally called "emblem glyphs" and attest to the presence of rulers, no doubt of varying strength, duration, and prestige, during the Classic period. Regardless of the controversy over whether or not emblem glyphs should be interpreted as expressing political autonomy or hierarchical relationships among polities (Marcus 1993; Martin and Grube 1995; Mathews, 1991; Stuart and Houston 1994), the types of interaction recorded hieroglyphically with emblem glyphs is revealing and includes warfare, marriage, heir designation, and royal visits (Martin and Grube 1995; Stuart and Houston 1994).

Returning to Formative period monumental architecture, the iconographic program—in the medium of stucco—executed on the front façades deserves special attention. As Freidel and Schele (1988) have noted, the iconographic programs of Formative pyramids are extremely depersonalized and rather deal with grand cosmological schemes and supernaturals such as the principal bird deity. Furthermore, pyramids do not seem to have been conceived as funerary shrines for dead dynasts until well into the Early Classic period. Sometime after AD 300, a fundamental shift occurred in the concept of pyramids; what had previously been culturally constructed "mountains of the gods" became *wits* for divine ancestors. As Hansen (1998) has indicated, the triadic pyramid complexes of the Late Formative do not contain royal tombs. Tombs of Formative construction that have been recorded in the Maya region often are discovered within modest constructions [e.g., Tikal Burial 85 (Coe and McGinn 1963); Los Mangales Middle Formative tomb (Sharer and Sedat 1987)].

While pyramids may define centers of power, pottery on the other hand can inform us about degrees of interaction (or isolation) of centers from each other. One of the most remarkable characteristics of the Late Formative period is the widespread manufacture of a type of pottery called Sierra Red. This so-called "Chicanel ceramic sphere," extending from Chiapas to Belize (roughly over 250,000 km^2), involved the construction of distinctive red monochrome bowls, dishes, plates, and spouted jars. Closer examination of Formative pottery from northern Belizean sites by Angelini (1997) (see also Bartlett and McAnany 2000) has revealed that apparent homogeneity in surface finish masks distinctive and locale-specific motifs, forms, and techniques of slip application. The identification of pottery with place, couched within a gloss of regional uniformity, suggests interaction between well-defined settlement cores characterized by distinct artisan traditions. In the terminal portion of the Formative period, a new style of pottery, mostly large serving bowls with overblown mammiform tetrapodal feet, swept through northern Belize and parts of the Petén, further emphasizing the tight interactivity of Formative period polities. Rather than being emblematic of a new class of kings, however, this pottery, with all of its allusions to fertility and its probable role in feasting, is simply added to the existing Chicanel assemblages. To summarize, interactive polities need not be assumed to have existed during the Formative period; they can be empirically demonstrated.

WARRIORS AND MILITARY ENGAGEMENTS

Without question, Classic Maya rulers were represented as warriors. Their military exploits (including the taking and sacrifice of captives) were amply recorded on murals and in stone. There are indications that warfare became increasingly common during the Late Classic period (Demarest 1992; Webster 1999:318), and it is true that military matters begin to vie with

INSTITUTIONALIZATION OF HIERARCHY IN THE MAYA REGION 137

parentage statements in the corpus of Late Classic texts. Iconographically, Late Classic rulers are more likely to be shown with a spear in their hand, while Early Classic rulers generally cradle in their arms a celestial bar of office (see Fig. 1). Around the end of the fourth century, Early Classic rulers are first imaged as warriors. Significantly, they are not shown with the traditional Maya long spear shaft but rather with the *atlatl*, war club, or Tlaloc shield of the central Mexican warriors [e.g., Curl Nose, 10th ruler of Tikal as shown on the side panels of Stela 31 (Fig. 4), and Smoking Frog, possible

Figure 4. Curl Nose, 10th ruler of Tikal as shown on the side panels of Stela 31 (drawing by J. A. Labadie after Jones and Satterthwaite 1982:Figs. 51 and 52).

usurper of the throne of Uaxactun, as shown on Uaxactun Stela 5]. On the other hand, there are no overwhelming indicators that Maya kingship was originally constituted around a notion of prowess in military affairs. Ample evidence of stemmed macroblade production from the Late Formative lithic debris piles of the stone-tool workshops at Colha (Shafer and Hester 1991) coupled with systems of Late Formative defensive ditches and walls at Becan, Edzna, and El Mirador suggest that armed conflict between polities has a strong Formative precedent.

If we admit the rather overwhelming evidence of a formalization of warfare in the fourth century while retaining the notion of a preexisting perhaps more *ad hoc* warrior tradition, then the codification of "leader as warrior" appears to postdate the emergence of kingship in the Maya lowlands. Even in the Late Classic, however, the frequency and importance of warfare may be overemphasized in current scholarship. At no point in time are Maya "warriors" ever buried with weapons in the manner that Kristiansen (Chapter 5, this volume) documents for the warrior aristocracies of the northern European Bronze Age. The lack of weapons in burial contexts suggests that ultimately the persona of a ruler as well as the perceived journey of the deceased into an afterlife did not include military combat. Furthermore, for every image of a Maya ruler holding a spear and standing on a captive there is another image (generally on polychrome vessels) of a ruler holding court, listening to petitioners, and receiving tribute (Stuart 1999:409–417). The sagacity of Solomon no doubt was as important as the sword of David. At the risk of confusing representation with reality, it does seem auspicious that the earliest Maya rulers are imaged as cloaked in ancestral heirlooms (particularly greenstone celts), cradling a celestial serpent bar, and engaging in actions that suggest a mediating role between humans and supernaturals.

NARRATIVES OF HEROISM

This factor is construed by Chang (1983) as that which gives a dynasty the moral authority to rule through reaching back into the past and linking current and aspiring rulers to past meritorious and exemplary deeds. Chang (1983:35) further suggests that the so-called collapse of Chinese civilization is simply the loss by political dynasties of their claim to moral authority. The extent to which Maya rulers relied on moral authority is unclear, but narratives of heroism certainly exist within the Maya prose tradition. The well-known highland Maya Quiche epic called the *Popol Vuh* is a Colonial period document in which the exploits of the hero twins, Hunahpu and Xbalanque, are recounted (Tedlock, 1985). Partly a morality tale and partly a recounting of genealogies, the *Popol Vuh*, which was written in Quiche Mayan using a European alphabet, is generally considered to have deep pre-Columbian roots. Central to the thread of this heroic narrative is the struggle of the hero twins to avenge the death of their father who was vanquished while playing the ball game against the Lords of the Underworld. Along the way, they also best the vainglorious 7-Macaw (representative of the dangers of self-magnification) and their evil stepbrothers. The athletic prowess of the hero twins is demonstrated in their eventual victory over the Lords of the Underworld in an epic ball game event. A few of the scenes painted on Classic period cylindrical vessels appear to relate episodes or variation on episodes from the *Popol Vuh,* which suggests that the Colonial Quiche document is one version of a corpus of very old narratives of creation and heroism. Furthermore, an Early Classic stela (No. 31) from Tikal shows the ruler as "Hun Ajaw," a Classic period variant of Hunahpu (Stephen Houston, personal communication, 1997). This depiction suggests that rulers may have been expected to embody the heroism and athleticism of the twins.

The heroism of the twins is based in part on their cleverness and athletic prowess. The very fact that they are twins sets them apart as iconic of principles of duality. Many societies have narratives of mythic heroism that feature the biological wonder and duality of twins; Romulus and Remus, the founders of Rome, easily come to mind. Hunahpu and Xbalanque, however, were ballplayers *par excellence* and battled the Nine Lords of the Night to avenge the death of their father. We now know that some form of ball game-based athleticism existed during the Early Formative period on the Gulf Coast as evidenced by the excavation of dense rubber balls at the Olmec water shrine of Laguna Manatí (Ortiz and Rodriguez 1994). Formative period ball courts at the Maya sites of Nakbe, Colha, Cerros, and Pacbitun further indicate ritualized athleticism within a formalized architectonic space prior to the emergence of kingship. Just as the oldest rubber balls in the world were linked with offerings to supernaturals at Laguna Manatí, so later Classic Maya iconography of ballplaying was linked with sacrifice and the afterlife (Miller and Houston 1987). This weighty conflation of ritual, athleticism, and competition may have constituted what Chang calls "moral authority" as rendered in a Mayan *oeuvre*.

WRITING AS A MEANS OF KINSHIP VALIDATION AND A KEY TO GOVERNANCE

In grappling with the fifth factor, we come face to face with the single most defining feature of Classic Maya kingship: A sophisticated and complex hieroglyphic script. This script, more than any other characteristic of Classic Maya society, reveals the extent to which access to knowledge was a prerequisite of political power. The emergence of what Chang (1983:90) has described as a "knowledge class" in ancient China bears strong parallels to the ancient Maya of the Classic period. Here again, we see the "bundled" nature of power in an archaic state wherein athleticism, writing, knowledge, wealth, and access to supernaturals can be perceived as different faces of power rather than as separate states that sometimes co-occur.

The four extant, pre-Columbian Maya screen-fold codices (the Madrid, Paris, Dresden, and Grolier) were painted on bark paper and reveal information of a divinatory and astronomical nature. From these codices, we can surmise that writing in Classic Maya society indicated access to sacred knowledge of the sky, the earth, and meteorological cycles. If the rituals described in texts carved on Classic sculpture are any indication, writing also indicated access to the wisdom of the ancestors and of supernaturals and perhaps even to a prestige language that differed from that of most commoners (Houston et al. 1997). Moreover, hieroglyphic texts carved on stone monuments revealed the relatedness of kings to their ancestors with a very indistinct boundary, if one at all, separating ancestors from supernaturals (Marcus 1992; Schele 1992).

The exclusionary knowledge of the elite class is emphasized in a passage in a Colonial period Yucatec text called the *Chilam Balam of Chumayel* (Roys 1967). Specifically, the text described a testing procedure whereby those who were pretenders to the throne, that is, not properly educated, would be tested and culled from the pool of potential rulers upon failure to understand a language called "Zuyua." Although there are many theories about the origins of Zuyua, one compelling idea suggests that Zuyua was based on the prestigious Cholti-related language used by Classic period elites (Stephen Houston, personal communication, 1997). This language would not have been generally intelligible to northern Yucatec Maya who had not been formally schooled in this archaic yet prestigious language. But members of this knowledge class were more than just top performers in a kind of college quiz bowl; they possessed a special kind of amalgamated power and knowledge that separated them from commoners. Perhaps the Classic Mayan term for vital power or force (*ip*) captures this quality much in the same way that Yoruban Ijesha characterize their divine kings as possessing *olaju,* an amalgam of knowledge and power (Feeley-Harnik 1985:292). Only those possessing such power are considered to be fit to govern.

But is not this knowledge class and its most powerful diacritical tool—hieroglyphic script—a result of kingship rather than a factor in its rise? Long before lowland Maya sculptors chiseled out long, double-column hieroglyphic texts linking rulers with the "First Mother," sculptors on the Gulf

Coast, the Chiapas and Maya highlands, and the Pacific Coast were hammering out narrative scenes showing supernaturals who were sometimes associated with long-count and sacred calendar round dates, often with minimal or no accompanying hieroglyphs (Clancy 1990). Some of these monuments (such as stela C from Tres Zapotes, stela 2 from Chiapa de Corzo, stela 10 from Kaminaljuyu, stela 1 from El Baúl, and stelae 2 and 5 from Abaj Takalik) reveal a deep concern with temporal matters and perhaps reflect the increasing politicization of time, or the crystallization of a temporal cosmological axis, as Helms (1999) has discussed. As mentioned above, an early long-count date from the Maya lowlands (stela 29 showing Scroll Ahau Jaguar of Tikal) is accompanied by a single-column long-count date with very little additional hieroglyphic text (see Fig. 1). The event depicted appears to be an accession to office; an ancestor (shown with head only) of Scroll Ahau Jaguar looks down on the ruler who cradles the doubled-headed serpent bar, emblematic of rulership. Indeed, the iconography is so saturated with the symbolism of inherited rule that accompanying text hardly seems necessary. In fact, it is only during the Late Classic period that stone sculpture as well as cylindrical drinking vessels become media for lengthy, hieroglyphic texts.

This trend of first pairing supernaturals with long-count dates (outside of the Maya lowlands) followed by the initial pairing of Maya rulers with long-count dates within the lowlands seems to confirm Helms' (1999) prediction that control of the temporal cosmological axis is essential to a concentration of power. Furthermore, pairing long-count dates with human rulers aided by ancestors can be viewed as effecting a rotation of the so-called temporal cosmological axis so as to situate a particular house at a point of origin or *axis mundi*. Only after this primacy is established and royal houses separated from those of commoners, do hieroglyphic scripts begin to play a role in diacritically affirming the deep chasm between rulers and the ruled.

EXCLUSIVE ACCESS TO ANCESTORS AND SUPERNATURALS THROUGH SHAMANISTIC RITUAL

Advancing to the sixth factor, we encounter what is perhaps the most contested and questioned domain of Classic Maya rulership: shamanistic practice. A shaman may throw him- or herself into a trance and pass over to the realm of the supernatural but Maya cosmology brought ancestors and supernaturals, into the realm of culture. Does this represent a fundamentally different approach or just a semantic variation? There are at least two poles of thought on shamanistic ritual in Maya society. One pole, most elegantly expressed by Freidel et al. (1993) in *Maya Cosmos* holds that Maya rulers derived a good deal of their power from serving as mediators between gods and humans and that this role was actualized through shamanistic practices such as bloodletting and transcendental visions. From this perspective, many of the practices and accoutrements of contemporary Mayan shamans, including the possession of an animal coessence, can be retrofitted onto Classic Maya royalty who represented a more elegant expression of basic cosmological precepts that continue to structure Maya thought and action today. The emergence of rulership in this perspective is not so much an ontogenetic process as it is a coupling of extant ritual with political power; ample "financial" backing served to elevate this dyad to a new level of shamanistic elaboration.

Another pole of thought is expressed by Houston and Stuart (1996) who focus not on shamanism per se but on the divine role of kings in summoning supernaturals (both ancestors and deities), physically housing and caring for icons of supernaturals, impersonating supernaturals, and taking on deity names and attributes. Of the last, perhaps that most clearly discernible is the term—*ch'ul* or *k'ul*—often glossed as "holy" or "divine" and scripted as the head of God C (Houston and Stuart 1996:292). This term is a vital prefix of the emblem glyph and so indicates a claim to divinity or ultimate authority by Classic period royals, who in effect assert that they are divine kings. The Mesoamerican belief that gods can take on a physical form and presence among the living has long been recognized among the Aztecs who codified it as *teixiptla,* meaning impersonator, image, or substitute (Boone 1994:105; Gruzinski 1989). A similar expression—*u-bah-il,* meaning "his body or image"—is found in Maya hieroglyphs and often is accompanied by the name of a deity and the identity of the impersonator (Houston and Stuart 1996:299). Images of Maya royals in the guise of the fire god, the sun god, or the maize deity suggest that this ritual "taking on" of the power—the *k'ul*—of Maya deities was another aspect of Maya kingship. A further link between rulers and gods was forged through the custodial responsibility accepted by rulers. That is, Maya rulers appear to have been responsible for safeguarding icons of the gods as well as sacred bundles. Such materializations of the gods provided potent symbols of group identity and the well-being of a group could depend on the careful handling of an icon of a tribal or ancestral deity. In the Mexican highlands, Aztec narratives recount a similar situation in which the safeguarding of the tribal icon, Huitzilopochtli, was of utmost importance, particularly during the long and perilous migration of the Mexica into the Basin of Mexico as well as their subsequent, largely successful, efforts at gaining political hegemony over competing polities within the Basin.

Schele and Miller (1986) have collected iconographic and glyphic evidence that gods and "tribal deities" were conjured or summoned by rulers with incense and blood-spattered burned paper. Such "auto-sacrifice" was undertaken to commemorate important (and probably infrequent) political events, such as an accession to the throne. The powerful visual images of Maya royals engaged in bloodletting indicates the significance and possibly the rarity of royal bloodletting. For comparison, Catholic iconography is replete with redundant images of a bleeding man hanging on a cross. Who, unfamiliar with Christian orthodoxy, would surmise that this event, iconic and central to Christianity, was a single and unique occurrence?

In the analysis of Maya rulers offered by Houston and Stuart (1996), rulers are seen to perform many religious duties that are analogous to shamans but within a much more structured and institutionalized context. Their authority stems not from their ability to conjure ancestors (as among contemporary Tzotzil shamans); rather, the conjuring of ancestors is an expression of their authority. To perform this duty well is to increase ones $k'ul$ or power possibly at the expense of another ruler (if power was conceived as a finite entity). Conversely, a poor ritual performance could indicate a seepage of divine power or a weakening of the divine power of the ruler much in the same way that Crumley (Chapter 2, this volume) argues that poor harvests weakened the power base of divine Roman emperors. In fact, Demarest (1992) has noted that the pulses of military victories and construction events at individual Classic Maya polities never appear to occur synchronously. Rather, as the fortunes of one polity wax, those of another wane.

The nomenclature and performance aspects of divine rulership, as reviewed above, certainly are results of a process rather than causal factors of an emergent process. They are paragraphs of a divine charter rather than a recipe for one. In reference to the Inca, the forging of a charter or political ideology has been described by Conrad (1992) as a propagandistic affair, the simplification of a complex cosmology with a resulting codification of an imperial ideology centered on Inti, the sun god, and the royal founding ancestors. Marcus (1992) too, has stressed the propaganda inherent in the written medium of Maya Classic texts. Can we discern a "great simplification" of Maya cosmology attendant on the emergence of kingship? Is "shamanistic politics" in effect a simplification of Formative period cosmology? Based on the foregoing discussion, I think not; rather, the opposite seems to have occurred, with an increasingly complex array of ancestors and supernaturals to be venerated, placated, summoned, and impersonated through increasingly elaborate ritual performance. The pageantry alone must have been convincing testament to the concentration of power and wealth within Classic period political capitals, which brings us to a consideration of our seventh and final factor.

WEALTH AND ITS AURA

According to traditional notions of power in Javanese society, wealth is a manifestation of power and not a route to it (Anderson 1972). We can follow this line of thought by noting that Chang (1983:8) asserts that civilization is a manifestation of accumulated wealth by which he is referring to the extraordinary concentration of scarce and precious raw materials, the elaborate fabrications of highly skilled artisans, and sometimes the massive granaries filled with crop harvests. It should be stressed that the characteristic of power and wealth of interest is its disproportionate distribution; in fact, its unmitigated concentration in the hands of a few. This process seems to get underway as soon as house-based societies develop. The fixity of place and the attendant changes in family labor structure engender profound transformations in ethical issues (such as principles of sharing), cosmology, and in the conveyance of material wealth transgenerationally. At this point, the door was opened to concentrations of power and wealth. Even though the Maya lowlands may not have been seriously populated before 1000 BC, by 400 BC indicators of disproportionate wealth distribution are present. Certain families were building bigger houses, directing the construction of massive pyramid complexes, and burying their dead with more lavish accoutrements. Grove and Gillespie (1992) suggest that across Mesoamerica in general the onset of the Middle Formative (ca. 1000 BC) marked an ideological transformation expressed materially in a decline in the use of pottery as a readily accessible medium of status in favor of more inaccessible "valuables" such as jadeite. In effect, a narrowing of the field occurred. By the end of the Formative, access to jadeite was not the issue, but access to heirloom-quality ornamentation as exemplified by the Dumbarton Oaks pectoral, stylistically of Olmec origins with the later addition on the reverse side of an etching of a Maya lord accompanied by an accession text (Schele and Miller 1986:119–120) (Fig. 5). Elsewhere, I have discussed the heirloom quality of Classic Maya elite ornamentation that was frequently carved with ancestor images (McAnany 1998). Jadeite is simply one of the most durable inherited heirlooms. Powerful but perishable sacred bundles with deep genealogical depth are a common motif within Maya iconography and a frequent touchstone in Colonial-period prose as well as contemporary Mayan ritual practice. The notion of inherited wealth—and all of the power and sacred knowledge that goes with it—is so central to Mayan thought that Yucatec Maya quickly adopted the Spanish practice of writing wills and last testaments (using Mayan scribes) during the Colonial period (Restall 1997).

Wealthy families may exist, however, in the absence of hereditary rulers, which leads us to the question of how disproportionate wealth trans-

INSTITUTIONALIZATION OF HIERARCHY IN THE MAYA REGION 145

Figure 5. The Dumbarton Oaks quartzite pectoral, stylistically of Olmec origins with the later addition of an etching of a Maya lord accompanied by an accession text (drawing by J. A. Labadie after Schele and Miller 1986:Plate 32–32c).

lates into kingly power. In many parts of the Maya area, evidence suggests that it simply did not translate; that is, wealthy kin groups and local leaders existed, but kingship, institutionalized rulership, never crystallized (McAnany 1995). Northern Belize is a part of the Maya lowlands that remained densely populated throughout the Classic period, and while pyramids were constructed and pottery styles indicate interaction with the heartland of kings in the central Petén, the materializations of kingship—hieroglyphic texts, palace structures, and lavish tombs—are rare to nonexistent. In the Petén and elsewhere, however, kingship emerged, apparently from the house compounds of wealthy families who probably laid claim to much of the best land of the surrounding region and no doubt claimed to have been among the "first founders" (see Schele 1992, for an explication of the founder's glyph in Late Classic Maya inscriptions). In the Maya lowlands, the "first founders" seemed to have taken up the task of "wrestling" with the cosmos and establishing a ritual *axis mundi,* which generally entailed the construction of a communal, if not monumental, facility for ritual practice. Evidence from the northern Petén Formative sites of El Mirador and Nakbe among others indicates a quick escalation of expectations regarding appropriate scale of ritual structures. Although controversy exists as to whether or not institutionalized leadership is a prerequisite of monumental construction (see Burger 1992; Billman, Chapter 9, this volume, for a contra viewpoint), nevertheless, large-scale monumental construction does require higher-level coordination, or leadership, and management of a sizable labor force. There probably is an inverse correlation between the number of workers needed to build a structure and the number of individuals who enjoy sanctioned access and use of structured sacred space. In the Petén, once competitive interaction took hold, then a premium was placed on showcasing the finest ritual structures, and leaders with the oldest pedigree (and associated heirlooms) and most venerated ancestors.

In its most fundamental form, the wealth of Maya society came from the earth, from the crops that were harvested. Depictions of the Maya earth monster portray a supernatural with a slightly frightening demeanor and none of the elegance of the young, handsome maize deity. As all farmers know, growing food is risky business and reducing that risk has been a main concern of farmers since the "taming" of food crops. Largely dependent on rainfall agriculture, Mayan farmers were a people who Netting (1993) would have referred to as "small holders." With fields dispersed in various wet and dry locales as well as next to the family compound, Maya farmers minimized their risk by maximizing the diversity of field locales rather than attempting to monopolize one particular edaphic niche. The centrifugal pull of this type of farming has long been noted, but the long-term (over 1500 years) productivity of this food production system cannot be denied. Netting (1993) would attribute this success to the small holders who, finely attuned to the subtle nuances of edaphics, slope, wind, and water, work their lands with a conservation ethic and knowledge that cannot be reproduced on large-scale estates. It is precisely this care and agricultural knowledge that yielded a wealth base sufficient to fuel the emergence and maintenance of kingship within the competitive political and ritual milieu of the Petén.

The large extended family residential compounds that work well with this type of farming are present in the lowlands by the Late Formative and persist, in many sectors, through the Colonial period. These corporate groupings with well-defined heads of household contained clear asymmetries in power and wealth (see McAnany 1995). They also were places of ritual practice, especially as related to ancestors, and repositories for ritual paraphernalia. In short, they contained many of the "preconditions" of rulership with some very key differences. What did it take for a powerful and wealthy head of household to extend his power base to an entire village and to

INSTITUTIONALIZATION OF HIERARCHY IN THE MAYA REGION *147*

transform that village into a central, political capital? We consider this question in the final section.

FINAL THOUGHTS ON THE RISE OF RULERSHIP IN THE MAYA LOWLANDS

Across the time–space continuum of the Maya lowlands, we can discern an early emphasis on ancestors who seem to have provided a link between the present and the cosmological "beyond and before" and whose continued presence within a "house" via subfloor burial clarified the resource rights and privileges of the resident kin group. Moreover, within the established and burgeoning communities of the Late Formative, demarcation of sacred space through the construction of monumental architecture indicates the emergence of a unified concept of ritual and political power. This unified concept is key to rulership and the centralization of power. Does this amalgam foreshadow the bestowal of kingly powers on village shamans or does it more likely presage a coopting of shamanic skills to be recombined in a potent mixture with cosmological connections to deified ancestors and ultimately to First Mother and First Father? The latter seems a more likely pathway to rulership. Emphasis during the Formative period on individual status, kin hierarchies with identified first founders, the temporal framework of ritual cycles, and the performance aspects of ritual practice appear to have laid the foundation for the emergence of the very personalized style of rulership that is so characteristic of the Maya lowlands.

Concentrations of wealth also must have played an important role in the transition to rulership. Kin groups physically constituted in large residential compounds facilitated such concentrations; moreover, these compounds served as the nexus of activities during cycles of feasting and ritual events. Physical heirlooms of inheritance, much more numerous and precious within wealthy households, would effectively and diacritically mark certain households as more strategically positioned and privileged, and thus able to play a greater role in the recombinant process of giving birth to rulership. A key feature in this process appears to have been a new type of ritual that stressed temporal connections to ancestral deities and required the formal construction of ritual space, desirably monumental in scale, to house icons of these deities and to enact rituals to ensure their continued favor.

Ritual space also is vulnerable space as it is often perceived as the soul of a people. As such, it requires special protection, and this responsibility, of course, falls to the powerful households from the ranks of which young men skilled in competitive ballplaying came forth to don the regalia of warriors. Although earlier Formative period hostilities between villages may have been quite common, warfare linked to ancestral deities was imbued with an additional potency. In effect, the stakes were raised; there was great power to be won or lost. Judging from the degree to which rulership was materialized iconographically, textually, and architecturally in the Maya lowlands, there seem to have been clear-cut winners and losers. In the end, one of the most striking characteristics of Maya society and Mesoamerican society in general is the masterful way in which a charter of centralized rulership was crafted from cosmological precepts rooted in nature. Commonly employed metaphors of kingship derived their inspiration from natural principles of hierarchy, symbiosis, regeneration, and continuity. This ideology left no stone unturned and managed to be all inclusive, "natural," and in ritual action so exemplary of cosmological order.

Part *IV*

Rulers in Power

The relationship between chiefdoms and states is a murky one in the archaeological literature. Different scholars define the terms in alternative and sometimes conflicting ways, and there is little consensus as to how chiefdoms and states are to be recognized in the material remains of the archaeological record. The preceding chapter by McAnany helps to highlight the kind of empirical, historical changes in power and authority that are manifest in the trajectory of political centralization as a system is transformed from chiefdom to state and leaders become rulers. The chapters in the present section look at leadership strategies and the exercise of power in state societies governed by rulers. Again, as will be seen, it would be a mistake to assume that power and political centralization in state societies is monolithic and concentrated in the hands of kings or lords. Just as in the present day, politics in archaic states were negotiated and there was variation as well as pattern in the tinkering of state rulers.

Like McAnany, Feinman (Chapter 8) also uses the example of the rulership of Maya kings, but in this case compares it with that of the ruling elite of the contemporaneous Teotihuacan polity in central Mexico. His analysis here, a natural progression of his previous work in Mesoamerica, North America, and more recently even China, is comparative and synthesizing. In Chapter 8, Feinman illustrates why it is important not to overgeneralize about the concept of "centralization" and assume that it means the same thing in all complex polities. By looking at iconography and other material remains, Feinman argues that the highly visible, grandiose Maya kings were pursuing one kind of rulership strategy—his "network" organization—while the much more faceless, bureaucratic lords of Teotihuacan were pursuing another—his "corporate" organization.

The origins and development of the extraordinary coastal polities on the coast of Peru provide the empirical foundation for Billman's (Chapter 9) analysis of the exercise of power by state rulers. In this case, he is trying to understand both why centralized political authority arises first in some places and then expands in other areas. In looking at the Peruvian coast, he makes two central and seemingly unrelated observations: northern valleys tend to have more arable land than do southern valleys, and El Niño (disruption of ocean currents and torrential rains) events occur more frequently in the north than they do in the south. He then makes the argument that the central coast is a kind of crossing ground where both the stimulus and the means for the exercise of centralized power come together. He then follows a persuasive line of reasoning to show how combined environmental and geographic variables have a significant impact on the level of power exercised by rulers in the evolution of states societies along the coast. His analysis sheds light on the specifics of why the Central Coast was a nexus of the earliest and some of the largest monumental architecture ever to appear in the Andes, and on general patterns of how rulers build their bases of exerting power over their populations.

In Chapter 10, Stein uses Mesopotamia—the crucible of the world's first states—as a forum to examine the complexities of power politics in archaic state societies. Few would argue that Mesopotamian states were characterized by highly centralized forms of government with extremely powerful kings who were able to command virtually absolute obedience from subject populations. However, Stein uses his own data from extensive research in Turkey and reconsideration of textual and archaeological evidence from the Mesopotamian heartlands to paint a much more complex picture of centralized and decentralized power relationships. This is the clearest example in the volume demonstrating that while there is a cross-cultural trajectory toward increasing political centralization as societies become more complex, this trajectory does not translate into the inexorable concentration of power in the hands of a single omnipotent ruler. This chapter also shows that while the political tinkering of rulers can have a huge impact on the society as a whole, other portions of the population also are experimenting with their roles as social, political, and economic actors. And this tinkering outside circle of the ruling elite creates the highly diverse range of political relationships that historically characterizes complex state level societies around the world.

Chapter **8**

Mesoamerican Political Complexity
The Corporate–Network Dimension

GARY M. FEINMAN

INTRODUCTION

In 1960, a group of renowned archaeologists met in Burg Wartenstein, Austria, to discuss the path of human history from the late Pleistocene to the threshold of urban civilizations. The symposium, which culminated in the important volume, *Courses toward Urban Life* (Braidwood and Willey 1962), helped set a foundation for the past four decades of archaeological research. Great empirical contributions have been made during this period (e.g., Meltzer et al. 1986). Nevertheless, although our archaeological database is far richer than it was 40 years ago, my optimism was tempered by the realization that many of the same key theoretical queries and major issues raised in the volume remain "in play" and "up for grabs" today.

For example, in Gordon Willey's (1962:101) chapter on Mesoamerica, he puzzled over the nature and the basis of the differences between lowland Classic Maya (AD 250–900) and contemporaneous highland Teotihuacan (AD 200–650) society and settlement. Four decades later, we still seem to

---

GARY M. FEINMAN • Department of Anthropology, The Field Museum, Chicago, Illinois 60605.

From Leaders to Rulers, edited by Jonathan Haas. Kluwer Academic/Plenum Publishers, New York, 2001.

151

lack the constructs and concepts to understand the nature of rulership and political organization in these ancient Mesoamerican civilizations. Although most neoevolutionary thinkers view the Maya polities as states (rather than less hierarchical chiefdoms), some would see them as simply "less complex" or less urban than central Mexican Teotihuacan and would ascribe the differences between these societies to the dimension of complexity. Yet how then would these same scholars use this envisioned "lower degree of complexity" to account for the abundance of long written texts at Maya sites, in contrast to their almost total absence at Teotihuacan (Cowgill 1992a:232, 1997)? Likewise, how can the adherents of a monolithic approach explain that named and ostentatiously bedecked Classic Maya rulers trumpeted their victories in life and were commemorated by elaborate funerary contexts at death, while at the same time rich burials seem to be rare and specific, easily definable rulers basically absent at the central Mexican city? From a monolithic or unilineal perspective, extensive writing, concentrated wealth, the glorification of individualized rulers, and stark status divisions in funerary behavior should be more (not less) prevalent at the metropole, Teotihuacan. Alternatively, must we simply chalk up these differences to unique cultural traditions that have no comparative lessons or theoretical implications beyond the world of ancient Mesoamerica?

In this chapter, I explore and contrast aspects of Classic Maya and contemporaneous Teotihuacan sociopolitical organization and rulership as a vehicle to introduce a comparative axis of variability (Fig. 1) that I see as cross-cutting or orthogonal to the familiar dimension of complexity or hierarchy. Apparent anomalies that arise from this pairwise Mesoamerican com-

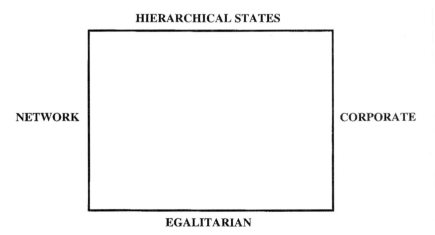

Figure 1. Comparative axes of variability between corporate and network modes.

parison serve as an entry point to challenge the traditional monolithic models of hierarchies, stratification, and power that are present in much anthropological thinking, pervading frameworks as diverse as cultural evolutionism and selectionism. More specifically, I challenge the notion that political hierarchy formation necessarily always entails the stark concentration of power and wealth in the hands of a very small number of individuals or specific families. I argue against the oft-held anthropological view that general political evolution or more hierarchical social formations can be measured by or equated with increasing centralization (e.g., Flannery 1972; Roscoe 1993). If centralization refers to the greater consolidation of wealth and power in the hands of a single overarching ruler, then larger, more hierarchical polities are not necessarily more centralized.

As an alternative to the more monolithic perspective on hierarchies, this chapter further defines, discusses, and develops a continuum between two modes of political organization. My colleagues and I (Blanton 1998; Blanton et al. 1996; Feinman 1995, 2000; Kowalewski 1998) previously have termed these the network or exclusionary mode on the one hand and the corporate mode on the other. Once introduced, these two different systems of political action are discussed in a comparative context. The exclusionary mode is shown to conform broadly to many current models of leadership and power that are in vogue in anthropology and archaeology today. While this leadership and organizational mode is recognized as often important and applicable, such as for the Classic period Maya, I propose that it is not uniformly or generally appropriate to all times and places. In fact, it is suggested that the failure to recognize the alternative corporate mode of organization has led to a number of apparent conundrums concerning several ancient political systems, most specifically, the ancient polity centered at Teotihuacan.

MONOLITHIC NOTIONS OF HIERARCHY

In archaeological models of political process and behavior, which otherwise reflect a diverse array of paradigmatic underpinnings, centralization often is directly linked with notions of sociocultural evolution and hierarchical development. I refer here to centralization in the political–economic sense of the concentration of power and/or the accumulation of wealth in the hands of a limited few. I am not using centralization as a synonym for settlement nucleation or aggregation, which represents a rather different (more demographic) use of the term.

The rather pervasive association of individual aggrandizement and power with political hierarchy is one that comes rather easily to Western social science. In the *International Encyclopedia of Social Sciences*, the concept of state is defined as "the geographically delimited segment of human society united by common obedience to a single sovereign" (Watkins 1968:150). From its origins, this normative concept of state has been closely linked to the concept of sovereignty, which was developed in the field of jurisprudence. This relationship stems from the basically legalistic assumption that all political societies are or ought to be united under a deterministic rule of law, and such a definitive law of the land cannot exist without a supreme lawmaking authority whose decisions are final.

This is not at all surprising, since the word "state" is said to have been fixed as a generic term for the body politic by the political philosopher Niccolò Machiavelli early in the sixteenth century, at which time it was used in the form *stato*. Slightly later, one of Europe's early modern statemakers, Louis XIV, is said to have remarked "L'état, c'est moi" (the state, it is me) (Fried 1968).

These roots of the state concept in Western thought clearly were influential to Fried (1967:237), who wrote that "the state must establish and maintain sovereignty, which may be considered the identification and monopoly of paramount control over a population and an area." More specifically, Ronald Cohen (1978:36) has stated that "(a) ruling class under a monarch . . . is a ubiquitous feature of early states." Likewise, Service (1971a) adopted a similar focus to leadership in his conception of chiefdoms. Almost by definition, chiefdoms have been viewed as the rule over a social group by one individual. For example, in a classic overview article, Carneiro (1981:45) wrote that "a chiefdom is an autonomous political unit comprising a number of villages or communities under the permanent control of a paramount unit." Of course, in each of these evolutionary schemes, chiefs lack the regular and institutionalized administrative structure typical of more hierarchical states.

The equation of increasing hierarchy with greater centralization was solidified in archaeological thought through Flannery's (1972) seminal paper, "The Cultural Evolution of Civilizations." Taking a systems approach, Flannery proposed centralization as one of the two core processes of cultural evolution. Although Flannery's own thinking on this topic has developed in some new directions (e.g., Marcus and Flannery 1996), his important paper has had great influence across the discipline. The association of hierarchy with the centralization of power and the consolidation of basic resources by a few individuals has been argued to underpin the emergence of the state in the conflict theoretical approach of Haas (1982:212–213). At the same time, it is a fundamental premise of Roscoe's (1993) practice approach.

Recent postprocessual and selectionist emphases on individual strategies have further promoted the presumption that highly centralized and personalized leadership, control, and aggrandizement underlie most if not all hierarchical political organizations. This notion is most pervasive in recent discussions of chiefly societies, where we see frequent references to accumulators (Hayden 1990), aggrandizers (Clark and Blake 1994), strivers (Maschner 1995), and entrepreneurial elites (Hayden 1995). Basically, these models link the processes behind chiefly emergence to individual self-interest and the strategies of accumulation and personal networking that are undertaken to foster that self-interest. Although the specific factors and the proposed sequences of change are not uniformly agreed on, each model views individual connections, status display, and unrestrained amassing of wealth as basic and generalized foundations to inequality and its institutionalization.

My argument here does not contest the viability of these models for the specific regions and cultural settings discussed by the aforementioned authors. Likewise, I certainly do not take issue with the theoretical consideration and importance of differential self-interest per se. Yet I do question whether self-interest is always manifested in the same way, and whether it is necessarily as blatantly enacted (in all spatiotemporal settings) as some of these networking models propose. Individual self-interest and the striving for achievement are not universally prevalent to the same degree in all cultural contexts as they generally are reconciled with and checked by more collective concerns. As a result, manifestations of individualistic behavior are in part a consequence of the larger social settings where they occur, and so the prevalence and materialization of such self-interest is neither uniform nor constant (e.g., Strathern 1981).

More to the point, a central premise of this chapter is that there is more than one strategy of political action or pathway to power, and that these different organizational modes would leave rather distinct archaeological correlates. Building factions, amassing wealth, and adorning and displaying oneself in an ostentatious way clearly are key themes in one pathway to power that I refer to as the network mode. But a key point here is that hierarchical organization also can take a somewhat more "faceless" and less centralized form. Furthermore, if we are to build convincing and broadly relevant comparative models, they cannot disregard important axes of variability. The next section briefly reviews the network and corporate modes.

THE CORPORATE AND NETWORK MODES

As defined here, the corporate and network modes (Blanton 1998; Blanton et al. 1996; Feinman 1995, 2000) parallel similar distinctions in leadership strategies and political economy that have been noted in earlier comparative analyses by Strathern (1978), Renfrew (1974), Drennan (1991), and others (Lehman, 1969) (Table 1). We have argued that these two general political–economic modes or paths of action (corporate-based and network-based) represent dual strategies of hierarchical organization. These modes are not proposed as reified or immutable societal types. Likewise, the corporate-based and network-based strategies are not proposed as mutually exclusive categories. In fact, to a degree these strategies coexist in the political dynamics of all social arrangements. As a consequence, empirical variation along the corporate–network dimension theoretically is continuous. Nevertheless, because these modes also are structurally antagonistic (see also Strathern 1969:42–47), they have been found to have had different degrees of relative importance cross-culturally. Their relative significance also may vary in a single region over time (Blanton et al. 1996). One or the other mode often dominates in any particular spatiotemporal setting (Blanton et al. 1996; Strathern 1969:42).

I explicitly do not wish to conflate corporate or network modes with specific cultural or ethnic "personalities." One cannot say that the Maya always followed network strategies, just as one would not characterize Maya society as always having been organized as states (e.g., Marcus 1995). The corporate and network strategies are neither static in a spatiotemporal sense nor are they culturally bound in a manner that implies that specific groups of people immutably follow specific strategies. Once network or corporate-based strategies are in place, one should not presume that the specific political–economic strategy necessarily will remain predominant or unchanging throughout a specific regional or cultural sequence.

For example, the ancient Romans had a republican government that eventually underwent key transitions. At a later date, during the Roman

Table 1. Political–Economic Typologies That Parallel the Corporate–Network Distinction

Network/exclusionary	Corporate	Reference
Individualizing chiefdoms	Group-oriented chiefdoms	Renfrew (1974), Drennen (1991)
Gumsa	*Gumlao*	Leach (1954), Friedman (1975)
Prestige goods systems	Big man competitive feasing	Friedman (1982)
Wealth based	Knowledge based	Lindstrom (1984)
Material based	Magical based	Harrison (1987)
Finance-based big man	Production-based big man	Strathern (1969)
Noncorporate organization	Corporate organization	Schneider et al. (1972)
Wealth finance	Staple finance	D'Altroy and Earle (1985)
Wealth distribution	Staple finance	Gilman (1987c)
Twem exchange	*Sem* exchange	Lederman (1986)

Imperial period, power was concentrated in the hands of specific emperors. Political strategies shifted greatly over these centuries, so Roman political behaviors as a whole cannot be categorically placed into either the corporate or network mode. Rather, it is the comparison across Roman history that is most intriguing, with two key indicators of a more network strategy—opulent ruler palaces and colossal portrait statues of the powerful emperors—first having appeared during the early Imperial period when a series of authoritative individuals ruled the ancient city (Reid 1997).

From a cross-cultural perspective, the corporate and network modes may have their basis in the external and internal components of the "dialectics of control" (see Giddens 1984:374; Spencer 1993). That is, those in leadership positions generally must balance what have been called "self-serving" and "system-serving" goals and interests. Although these dual concerns often have been proposed as opposing functionalist versus selectionist models, or as representing different stages in the stepladder of cultural evolution, both self-serving and system-serving considerations generally are at work in most hierarchical arrangements. Aspiring leaders who address and resolve the wants and needs of followers (larger societal concerns) simply are more apt to be retained and supported by factions despite other potentially selfish aims and strategies of those individuals.

In an important argument that foreshadowed our work (Blanton et al. 1996), Renfrew (1974) contrasted individualizing and group-oriented forms of chiefdoms. The cases that he outlined were recognized as relatively similar in overall sociopolitical complexity, yet organized in markedly different ways (Renfrew 1974:74). For Renfrew (1974:74–79, 83), group-oriented polities deemphasized differentials in access to personal wealth, while placing great importance on communal activities and group rituals, which link the geographic segments that comprise these populations. Collective labor and monumental public architecture were evidenced, while great stores and individual displays of personal or portable wealth were rarely apparent. Neolithic Malta, the henge-building peoples of the European third millennium BC, and ethnographic Polynesia were interpreted as examples of these group-oriented arrangements.

The emphasis was entirely different in Renfrew's individualizing forms, like those of Minoan–Mycenaean Greece (Renfrew 1974:74–79, 84). The egalitarian ethos was overcome as specific individuals were differentiated and privileged. Marked disparity in personal possessions existed, and the residences and tombs of rulers were far grander than those of the bulk of the population. Exotic trade wealth appears to have had a significant role in prestige accumulation. Some of the goods that were exchanged were elaborately crafted by specialists. In contrast to the group-oriented formations, communal ritual and public construction appear to have had lesser roles.

Significantly, in a more recent comparison of pre-Hispanic chiefdoms in the Americas, Drennan (1991:283) found a similar distinction. In two cases (central Panama and Alta Magdalena) from the Intermediate Area (lower Central America and northern South America), almost all public building was focused on individuals. Status differentiation was conspicuous. In the best documented of these cases (central Panama), tremendous volumes of portable wealth, reflecting great craftsmanship, were found in select burials. The items recovered also signal the importance of extraregional exchange links. Many of the burial goods evidently were meant for elaborate personal adornment. "A large portion of the resources mobilized in the local economies in central Panama and Alta Magdalena went into competition over status dominance, focused heavily on the person of the chief" (Drennan 1991:283). These cases illustrate clear parallels with Renfrew's individualizing chiefdoms and with what we call the network strategy.

In contrast, a markedly different organizational mode was evident in two Early-Middle Formative period Mesoamerican populations (Valley of Oaxaca, Basin of Mexico). In these regions, collective works primarily were carried out to construct public spaces (plazas and mound groups) for communal ritual. Although inequalities existed (as they did in Renfrew's group-oriented chiefdoms), they were not expressed elaborately in house construction or burials. For example, at this time in the Valley of Oaxaca, all houses were made of comparable materials (earth and cane), and the most elaborate burials were simply lined by stones and contained no more than a few exotic offerings. Craft specialization, long-distance exchange, and status competition were not entirely insignificant in these highland Mesoamerican regions during this period (e.g., Feinman 1991; Marcus and Flannery 1996). Yet I concur completely with Drennan's (1991:283) perception that these activities consumed a relatively smaller proportion of the "chiefly domestic product" as compared to the aforementioned cases from the Intermediate Area. These Formative era Mesoamerican cases illustrate elements of both the network and corporate strategies (see Drennan 1991); however, considerable emphasis is on the latter mode, particularly when comparisons are made with the chiefdoms of the Intermediate Area.

Focused primarily on these archaeological examples from Europe and the ancient Americas, both Renfrew (1974:74,82–85) and Drennan (1991:284–285) have argued that long-term social change in chiefdoms has followed distinct paths, which have involved somewhat different political–economic strategies and processes. Yet these different modes or strategies also are evidenced in other societal settings as well. In comparative analyses of nonstratified societies in the New Guinea Highlands by Strathern, two organizational strategies (home finance and home production) also are outlined (Strathern 1969, 1978). Home finance defines a mode in which the relative importance of big men reflects their position in a network of financial arrangements (Strathern 1969:65). Through the external exchange of portable wealth items (e.g., shell valuables, live pigs), obligations and alliances are created (Strathern, 1978:75). Big men are able to solidify their factions through the local distribution of long-distance gifts that are derived from trade partners. As described for the Melpa of Mount Hagen and the Enga, this mode bears striking resemblance to the strategies employed by the Panamanian chiefs (see also Helms 1979). The acquisition of wealth through individual, entrepreneurial linkages and the use of that wealth by these charismatic figures to attract factions defines our network-based mode.

With home production, a big man's prestige depends on the labor force of his own kin group or settlement to raise the goods needed for feasting and exchange (Strathern 1969:42). For household or settlement heads, access to labor and land is critical and is derived largely through corporate descent. As described for the Maring and Siane, the eminence of a big man is achieved by obtaining more land and labor than others (Strathern 1969:65). Siane big men do not accumulate great personal wealth (Strathern, 1969:50). Intergroup ties, portable wealth, and financial maneuverings are far less important with home production than they are with home finance (Strathern 1978:98–99). In Siane communities, corporate groups are integrated by initiation rituals that crosscut component social segments. Such ceremonies are described as much less significant among the Enga and the Melpa (Strathern 1969:44). Home production fits comfortably within the corporate-based mode (Blanton et al. 1996), with its emphases on food production, land, labor, corporate kin relations, societal segments, and the integration of those segments through crosscutting integratory social and ritual mechanisms.

The corporate–network continuum bears some similarity to the distinction previously drawn between societies having simultaneous (network) and sequential–ritual (corporate) hierarchies (Johnson 1982). Of course, Johnson's (1982) discussion differs in its narrower concern with decision making, avoiding direct tie-ins with economic considerations. Simultaneous hierarchies are defined as social arrangements in which a few central individuals exercise integration and control over a larger population (Johnson 1982:403). Alternatively, sequential–ritual hierarchies are described as relatively more consensual and egalitarian formations in which ritual and elaborate ceremonies often are argued to have key roles in the integration of modular social segments (Johnson 1982:405, 1989:378–381).

However, in contrast to Johnson's seminal discussion regarding sequential–simultaneous hierarchies, I argue that corporate strategies are not restricted to nonstratified (egalitarian in the sense of Fried) social formations. More specifically, this discussion challenges the unilinear proposition (Spencer 1993) that cultural evolution necessarily moves either simply and directly from more consensual to simultaneous (individually focused) forms of organization or toward greater centralization. Rather, the corporate and network strategies are presented as organizational modes that, to a great degree, crosscut societal variation in relative hierarchical complexity and stratification. One might view the corporate-to-network continuum as a dimension for the comparison of political economies that runs perpendicular or orthogonally to the long-recognized axis of hierarchical complexity. That is, one can have societies organized either corporately or in more network fashion at each different degree or level of hierarchical complexity.

The corporate–network distinction also parallels D'Altroy and Earle's (1985) contrast between staple and wealth finance. Yet our distinction transcends the economy to strategies of rulership and political process in a way that their useful dichotomy does not.

To synthesize this far-ranging discussion, the corporate mode emphasizes communal ritual, public construction, food production, large cooperative labor tasks, power sharing, nonlinear inheritance of rulership, social segments that are woven together through broad integrative ritual and ideological means, an emphasis on offices rather than personalized officeholders, and suppressed economic differentiation. Despite the presence of large architectural spaces, Renfrew (1974:79) aptly described individuals in such polities as relatively "faceless" and "anonymous." In contrast, the network mode places greatest significance on personal prestige, wealth, power accumulation, elite aggrandizement, highly individualized leadership (and its representation), lineal patterns of inheritance and descent, personal networks, prestige goods exchange, exotic wealth, princely burials, and the specialized manufacture of status-related craft goods (Table 2).

Table 2. Tendencies of Corporate–Network Modes

Network	Corporate
Concentrated wealth	More even wealth distribution
Individual power	Shared power arrangements
Ostentatious consumption	More balanced accumulation
Prestige goods	Control of knowledge, cogntive codes
Patron–client factions	Corporate labor systems
Attached specialization	Emphasis on food production
Wealth finance	Staple finance
Princely burials	Monumental ritual spaces
Lineal kinship systems	Secmental organizations
Power inherited through personal glorification	Power embedded in group association/ affiliation
Ostentatious elite adornment	Symbols of office
Personal glorification	Broad concerns with fertility, rain

MESOAMERICAN POLITICAL COMPLEXITY   *161*

THE CLASSIC MAYA AND TEOTIHUACAN

Mesoamerican archaeologists long have puzzled over the societal differences between the Classic Maya states (Chapter 7, this volume) and the polity centered at Teotihuacan (e.g., Willey, 1962) (Fig. 2). In the past, some scholars have viewed these differences as primarily reflecting the degree of "organizational advancement," with Maya polities seen as small and less complex than those at Teotihuacan (Sanders and Price 1968:139–145). More recent perspectives have tended to regard many elements of this variation as reflecting the kind, not simply the level, of organization (e.g., Blanton et al. 1993; Freidel 1981; Marcus 1983; Sanders 1981). Correspondingly, current epigraphic and archaeological interpretations (Marcus 1993; Martin and Grube 1995) indicate that some Classic Maya polities were both large and powerful, capable of exerting influence at other sites that were situated many kilometers away across large swaths of the eastern lowlands.

Nevertheless, a comparison of these basically contemporaneous polities indicates that Teotihuacan was more populous, more densely occupied, and more monumental architecturally than the Maya cities. Traditional models of cultural evolution that feature centralization as the key dimension might lead to the expectation that greater stratification of wealth and more differentiated burial contexts should be found in the former center. Likewise, based on these models, one might predict more concentrated power in the hands of a few select rulers at Teotihuacan as well. Yet an examination of the empirical record indicates the complete opposite. In the following discussion, I begin with a short outline of Classic Maya rule, funerary custom, iconography, and socioeconomy. The treatment is brief, since in many ways the Maya conform to our rather monolithic and ruler-centric conceptions of power. Subsequently, contrasts are offered with Teotihuacan, which appears to have had a starkly different organizational mode.

During the Classic period, the eastern or Maya Lowlands of Mesoamerica shared a single cultural, economic, and linguistic system but was subdivided into a series of polities. Many of these polities were small and autonomous; however, larger entities and federations were formed periodically through political alliance, elite intermarriage, and military conquest (Marcus 1993; Martin and Grube 1995). Lengthy texts found at Maya centers describe the exploits of named Maya rulers. Many of these texts were found on large carved stone stela that also fully depicted the ruling Maya lords in elaborate costumes that featured ornate feathered headdresses, cut marine shell belts, and carved stone jewelry (Schele and Miller 1986). At specific sites (like Palenque), distinctive headgear serve as crowns, which denote the wearer as ruler (Schele and Miller 1986:113–114).

Erection of such stela was a repeated and important political/ritual act.

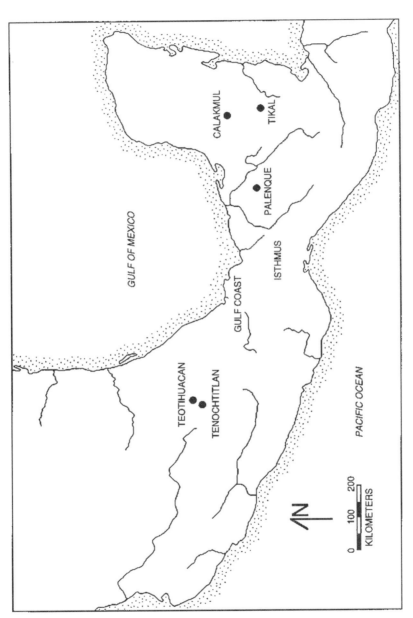

Figure 2. Map of Maya and Teotihuacan population centers.

Recurrent themes of these texts included the genealogical and marriage ties of rulers, their military exploits, and the rituals associated with their accession to rule. Decades of epigraphic research have enabled the reconstruction of dynastic histories for certain sites that can be traced and recorded for centuries (Culbert 1988, 1991).

Although Maya lords were legitimized in part through the rhetoric of royal descent, they also served as the dynamic mediators linking the natural and supernatural worlds through the veneration of their forebears. Mesoamerican peoples (in some regions to the present day) viewed massive tropical trees, like the ceiba, as the *axis mundi* (the world axis) that maintained cosmic order by separating the earth and the sky and the underworld (Kolata 1984). As a connector of vertical levels that maintained cosmic separations and so world order, the "world tree" (*axis mundi*) was a powerful symbol appropriated by various Mesoamerican polities (Gillespie 1993). Yet it was the Classic Maya rulers, in contrast to other Mesoamerican peoples, who actually had themselves individually depicted in art and visually referred to in texts as great trees (Freidel and Schele 1988). The artworks suggest that Maya rulers actually personified the world tree and that their bodies became a prime conduit for supernatural forces, perhaps during rituals (Gillespie 1993; Schele and Miller 1986). This exclusionary control of supernatural forces (as the key node that both linked and maintained the order between the levels of the cosmos) was an ideological basis of their power (Schele and Miller 1986). In contrast, later Aztec rulers, for example, occasionally were mentioned in written documents as having been like trees. The use was more metaphorical, and they themselves were not equated directly with the *axis mundi* (Gillespie 1993).

Maya rulers often were commemorated lavishly at death, buried under massive pyramidal mounds, as at Tikal (Haviland 1992; Jones 1991) and Palenque (Ruz Lhuiller 1973). Other marked disparities in funerary wealth and context also have been reported with elites frequently buried in painted tombs with jade, shell, polychrome pottery, and carved bone as well as other goods (e.g., Rathje 1973, 1980; Sharer 1994:163–164). Some of the most elaborate tombs included carved stone panels and could only be reached by underground stairways. These elite graves contrast with many interments that have been found in more mundane domestic contexts with few accompanying artifacts.

To procure adornments and funerary objects, Maya lords sponsored craft specialists and promoted luxury trade (McAnany 1993). These economic specialists often resided in elite households, producing items such as polychrome pottery that specifically glorified and was commissioned by their elite patrons (Reents-Budet 1994; Stuart 1988). There is increasing evidence that some of these craft workers were themselves elite, related to the lords that they served (McAnany 1993). Certainly I am not suggesting that the Classic Maya political economy was based exclusively on the production and exchange of exotic wealth items. Yet such goods were visible, culturally important, and unequally distributed. The production and control of wealth items and their distribution does appear to be one important basis of power among the Classic Maya.

At most Classic Maya centers, archaeologists have identified palaces and elite compounds that clearly are much more elaborate and different in layout than the much smaller, more modest patio groups where most Maya lived. At Palenque, the so-called Great Palace sits on a 10-m high platform at the core of the site. The platform, which is 100 m by 80 m in extent, is composed of a series of rooms and galleries arranged around several interior patios or courtyards. The masonry compound includes a steambath and three latrines and can only be reached by a stairway that ascends one side of the platform (Sharer 1994:288–289). Similarly, at Calakmul, Structure III, "the Lundell Palace," also was built on a raised platform. It is composed of 12 rooms with associated corridors. The palace compound included an elaborate subterranean tomb with spectacular offerings (Folan et al. 1995b). In contrast, most nonelite Classic Maya lived in small residential structures constructed partially of perishable materials, such as adobe and/or thatch (Sharer 1994:473–474).

Classic Maya society was marked by rather clear-cut distinctions in access, wealth, and power. At the core of these polities were specific elite individuals who used their personal networks of ancestors, affines, exchange partners, and political allies as a basis of their power. Portable wealth, ostentatious adornment, prestige goods exchange, attached specialists, ancestor veneration, the rhetoric of princely descent, the glorification and depiction of specific individuals, crowning headgear, and elaborate burials were all key features of this organization that was outlined above as network or exclusionary in form (Table 2).

Although the characteristics of the network or exclusionary mode broadly correspond to more traditional (monolithic) notions of complex political economies and leader-centric strategies of rule, that is not the case at Teotihuacan. As Cowgill (1992b:207–208) has written: "the office of rulership itself, or headship of the state, has proven curiously elusive at Teotihuacan, in contrast to its high visibility in Classic period Maya inscriptions and monuments." In fact, one is hard pressed to identify a single ruler at Classic period Teotihuacan in burial, text, or graphic depiction.

Teotihuacan art includes various glyphs and standardized signs, such as the "speech scroll" to signify speech. No lengthy texts, like those of the Classic Maya, are known to exist (see Cowgill 1997:135). As an analogy, many of the glyphs on Teotihuacan painted murals, incised ceramics, and carved stone panels resemble the standardized signs that can be seen today at interstate exits: the "gasoline pump" sign, the "plate and cutlery" sign, the "garbage can" sign, and the "dripping faucet" sign (Cowgill 1992a). As a consequence, we lack not only named rulers (Cowgill 1997; Hassig 1992:198), but recorded reckoning of descent and marriage relations, military exploits, or accession events as we have for the Maya. Teotihuacanos certainly were familiar with writing. They traded with the Classic Maya, and a Zapotec glyph has been found in the so-called Oaxaca Barrio at Teotihuacan. The various media of Mesoamerican writing (stone, ceramics, carved bone, painted murals) were as available to (and used by) Teotihuacanos as they were elsewhere in Mesoamerica. So the absence of lengthy texts would seem to be more of a cultural decision, strategy, or practice rather than a depositional circumstance or the mere lack of a key innovation. In addition, the largest collection of Teotihuacan signs, which recently were found painted on the floor of a patio (Cowgill 1997:135), is unaccompanied by any pictorial representation (human or otherwise).

Teotihuacan art is notable for the absence of scenes in which certain humans appear subordinated to other persons (Cowgill 1997:136; Millon 1988). In fact, much Teotihuacan art depicts signs and animals rather than individuals. When humans are present, they tend to be depersonalized and unnamed (Cowgill 1997:136; Pasztory 1992:292). In these representations, the people do not seem to portray specific individuals so much as people portraying certain roles. Although symbols of authority, such as special headdresses and flaming torches, have been recognized, they were not associated with specific or named personages. Rather, the various people who wore the tassel headdresses common to Teotihuacan art have been argued to belong to corporate organizations and to have held specific military and diplomatic offices (Millon 1973). Likewise, when seemingly well-dressed figures are represented in Teotihuacan art, they frequently are portrayed in groups, set in profile, dressed in similar attire, as though they were a "committee" or "council" for which their individuality was not essential (Pasztory 1997:56). In fact, their faces often are obscured. The "emphasis is on acts rather than actors" (Cowgill 1997:137). Many scenes and symbols recur over and over again in the art of Teotihuacan. Cowgill (1997:137) proposes the presence of ethos in which individuals were interchangeable and replaceable.

For Pasztory (1992:294), the famous (and rather standardized) Teotihuacan mask (found in greater abundance than masks at other Mesoamerica sites) is a prime example of the collective, nonindividualized aspect of Teotihuacan society. Given Steiner's (1990) explicit suggestions concerning the relationship between forms of leadership and specific kinds of adornment, the prevalence of masks at Teotihuacan as compared to the presence of unique crowns and elaborate headdresses among the Classic Maya would seem to be important. In fact, Steiner (1990) associates crowns with more egocentric and personalized strategies of rule and masks with more collective approaches (see also Ottenberg 1982).

Correspondingly, there are no definitive royal burials at Teotihuacan. No Classic period princely tombs have been isolated despite generations of fieldwork. Furthermore, in later Teotihuacan, the great majority of inhabitants lived in substantial, multiroom, multihousehold compounds that all approximated the canonical Teotihuacan orientation of 15.5 degrees east of true north (Cowgill 1997:137). These apartment compounds were internally variable; some rooms were more elaborate than others. In addition, some apartment compounds were larger and more finely built and decorated than other compounds. Yet they all shared a basically similar configuration, each with its own internal ritual space.

Despite architectural construction unprecedented in monumentality for ancient Mesoamerica, no definitive palaces or kingly residences with clearly specialized functions have been identified in the city (Cowgill 1983; cf. Flannery 1998). At Teotihuacan, the centrally located Ciudadela apartment compounds have been interpreted as residences of the heads of the state (Cowgill 1983). Yet the Ciudadela is unlike many better-known royal palaces, such as those in Aztec Tenochtitlán when the Spanish arrived or those of the Classic Maya (Cowgill 1997:151). To begin with, in the Ciudadela, there seems to have been relatively little differentiation among the component apartments.

A comparative analysis of burial contexts from three Teotihuacan apartment compounds has indicated that the differences in grave goods between burials within compounds were greater than the differences between the burials patterns in different compounds (Sempowski 1987; Sempowski and Spence 1994; see also Manzanilla 1997a). Each compound included interments with both a wealth and sparsity of offerings. As a consequence, although comparatively minor wealth differences do seem to be present between compounds, one cannot assign different compounds to distinct classes. Based on burial analyses, Sempowski (1992:52) concludes that accessibility to probable luxury goods and exotics, like jade, marine shell, mica, "does not seem to have been strictly controlled; rather the goods appear to have been available for use as potential grave offerings by groups of even quite average status at Teotihuacan." Various distributional studies of wealth and status items at Teotihuacan also indicate that exotic and highly crafted goods were not limited to the central part of the ancient city or any specific sector of the site (Cowgill 1992b:217).

Pasztory (1988, 1992) further argues that the art of Teotihuacan proclaims homogeneity and collective values in a society that was numerically large and clearly heterogeneous, having distinct ethnic barrios (Cowgill 1992b). Different hunting techniques (Manzanilla 1997a:117) and funerary customs (Sempowski and Spence 1994:265) were practiced in different apartment complexes. Yet, across the site, certain artistic and architectural conventions (e.g., *talud-tablero* architecture, Teotihuacan masks, three-temple complexes, tassel headdresses, a particular style of incense burner, moldmade figurines, specific glyphic representations, particular geometric designs) were highly uniform (Pasztory 1978, 1992). Millon (1992:389) has argued that in such a large city (with many inmigrants and ethnic diversity) social cohesion may have been fostered through this widespread ideology and the set of shared and standardized symbols.

At Teotihuacan, cosmological principles that linked rain, earth, sun, moon, and serpents with renewal and fertility were emphasized (Pasztory 1992:311–313). A recurrent theme in mural art represents processions of ritual officials or priests scattering seeds, corn cobs, and other food items that would have been associated with bounty (Manzanilla 1997b). Marine shells, a material generally linked to fertility in Mesoamerica, are prominent in the art and iconography of Teotihuacan (Kolb 1987). Collective rituals and symbolism associated with rain and fertility are a common feature of the corporate mode (Table 2). Although there is much that we still do not understand about the organization of Teotihuacan, strong corporate codes, redistributive rituals associated with fertility, and a segmental aspect that included the multihousehold apartment compounds and a quadripartite city plan were key features.

I should reaffirm that I am not asserting that urban Teotihuacan was free of rulers, completely devoid of wealth differentials, or that it was an egalitarian society in the sense of Fried (1967). In fact, it is conceivable that early Teotihuacan had a somewhat more ruler-centric organization (Cabrera Castro et al. 1991; Cowgill 1992c, 1997) than that found later in the Classic period (the focus here). The point of this comparison with the Maya is to suggest that leadership and political economy were organized differently in Classic period Teotihuacan than for the contemporaneous Maya states. I view later (post-AD 200) Teotihuacan as a hierarchical state, but a corporately organized one that lacked highly personalized rulership or blatant displays of wealth distinctions. Individual wealth and power were neither glorified nor emphasized. As Millon (1992:340) has observed, "rulership for the last 400–500 years of [Teotihuacan's] history appears to have been kept in check by a collective leadership." Cowgill (1997:152) has surmised that "supreme political authority may not always have been strongly concentrated in a single person or lineage." According to historic reconstructions (Paulinyi 1981), corulership and power sharing were by no means rare in post-Teotihuacan central Mexico, so there would seem to be some later analogical evidence for key aspects of the more collective leadership pattern proposed here.

REFLECTIONS ON MESOAMERICAN CALENDRICS

Before considering broad comparative implications of the corporate–network dimension, I offer a brief consideration of ancient Mesoamerican calendrics. I take this deviation because I believe that it illustrates the potentially important analytical avenues that may be opened for Mesoamerican archaeology by a consideration of the corporate–network dimension. One test of the efficacy of new ideas and perspectives is whether they help to unravel old debates. I also consider this calendric example because it serves to illustrate how different political–economic strategies affect the materialization of other facets of life even within a specific cultural setting.

At the time of Spanish Conquest, most Mesoamerican peoples used the Calendar Round, a time-keeping system that accounted for periods no longer than 52 years. The Aztecs, for example, conceived time as an endless succession of these 52-year periods. This cyclic conception of time was broadly shared in Mesoamerica, extending back to 500 BC, if not somewhat before.

When the Spanish arrived, the sole exception to this pattern was the Maya, who used the so-called Short Count, a temporal cycle that endured 256 years. Yet, for at least a century, epigraphers and archaeologists have known that the Classic Maya, the antecedents of those who used the Short Count, were able to keep track of much longer temporal episodes. The Classic Maya achieved this through their use of the Long Count, a time-keeping system for which no records exist following 909 AD (Sharer 1994:567-568). The demise of the Long Count basically coincides with the end of the Maya Classic period and a major reorganization of Maya society and political organization that took place at that time.

In contrast to the 52-year calendar round, the Long Count measures time from a fixed point in the deep past rather than in short repetitive cycles. In this aspect, the Maya Long Count in conception resembles the contemporary Gregorian calendar (which of course starts with the fixed date of Christ's birth). Because the Maya Long Count (like our own Gregorian calendar) is able to track thousands of years, it has a decidedly linear aspect, as it is able to give each day in a long span of time its own unique designation.

Since its recognition and decipherment, scholars have puzzled over the history of the Long Count. Our earliest records of the Long Count are found on carved stones in the Isthmus and Gulf Coast regions of Mesoamerica during the last century BC Shortly thereafter, this calendar system was dropped in those regions and adopted by the Maya. The Maya employed the Long Count on their stela and other texts from roughly the third through the tenth century AD.

Use of the Long Count by the Maya basically brackets the Classic period. Scholars long have wondered why this fixed point or syntagmatic calendar system went out of use at the end of the Classic period and why it was not employed more broadly across pre-Hispanic Mesoamerica. It is now evident that the end of the Maya Classic period was not marked by a complete regional abandonment or population replacement (Marcus 1993, 1998; Rice 1986). Yet it was characterized by a shifting of demographic and political power across the lowland region (with some net population loss), a curtailment of stela erection, a decline in the frequency of the ruler depictions, and the loss of the Long Count. Was the timing of this calendric shift simply happenstance or did the way in which time was kept and history reckoned change in concert with the transitions in architecture and political organization that characterized the later Maya Postclassic period (Sabloff and Rathje 1975; Sharer 1994; Stuart 1993:346–348)?

Perhaps we can gain perspective on these questions through a consideration of the long Roman sequence noted briefly above. In Republican Rome, a cyclical or paradigmatic conception of time was employed in which there was no fixed point (Walters 1996). Time was denoted through association with the ruling status of specific consuls, who composed the republican oligarchy. Yet new consuls often adopted the names of their predecessors, so different temporal eras could be associated with the same recycled consular names. This cyclical, repetitive aspect to the tracking of time has posed challenges to subsequent Roman historians, who wish to construct a linear timetable for the pre-Imperial era (Walters 1996:72–73).

The melding of the past with the present was evidenced further through the wearing of ritual death masks in funerary contexts, in which current consuls each wore the masks of their most renowned ancestors. In these events, dead ancestors who lived in different eras separated by time were then united. Through the wearing of masks, the living were assimilated with the dead. "The discreteness and sequentiality of history were effaced in this ritual, while uniqueness and individuality were replaced with similarity" (Walters 1996:84). In other words, the death mask rituals served to establish a sense of likeness and continuity across many generations, a conception reinforced by the paradigmatic system of temporal recording.

With the fall of the Republic and the institution of monarchy, the paradigmatic view of time and history also began to fade. This transition was marked by a shift from the hereditary oligarchy of republican families, whose collective careers endured centuries, to a monarchy, whose dynasties were relatively short-lived. The rule of unique, often idiosyncratic individuals supplanted the power and authority of collective lineages. Just as government, politics, and societal structure were altered, so too was the old view of history and of the past (Walters 1996:90). The use of the funerary masks and their associated rituals also shifted and declined, becoming more personal and private in nature (Walters 1996:85).

Walters (1996) postulates that the advent of a fixed-point Roman calendar with the establishment of a new order would have facilitated the tracking of specific royal genealogies and inheritance patterns as well as establishing a chronological basis for historical causality in regard to change"

> The essence of the paradigmatic viewpoint for history is to arrest change by reducing difference and diversity to similarity. The past, the present, and the future are made copies of each other. The syntagmatic method, on the other hand, admits the flow of history but controls its data by organizing its chains and sequences, whose connectors are causes. (Walters 1996:96)

In considering this example, parallels can be seen for the Maya and prehispanic Mesoamerica. The general adherence to a cyclical notion of time in ancient Mesoamerica corresponds to the prevalence of corporate polities in this world (Blanton et al. 1996). By contrast, the heavy use of the Long Count in the Classic Maya world probably served and enhanced their highly exclusionary political organization and the messages that their rulers wished to convey. As this ruler-focused structure with its attendant stela cult collapsed and more corporate polities formed during the Maya Postclassic, the corresponding changeover to a more cyclical (shorter cycle) calendric system seems to have been a consequence that serves as a converse to the Republican-Imperial transition described for ancient Rome.

THE CORPORATE MODE: COMPARATIVE UTILITY

To this point, I have discussed how the corporate–network dimension can provide useful avenues to understand variation and change in pre-Hispanic Mesoamerica. Yet these concepts are meant to have broader comparative utility. In this section, several examples are considered in which the concept of corporate hierarchies is illustrated to have analytical utility.

In several global regions beyond Mesoamerica and Teotihuacan, the traditionally presumed conflation between hierarchy and the centralized embodiment of wealth and power in specific individuals has led to murky debates and apparent enigmas. For example, scholars have long puzzled over the nature of rule and political organization in the ancient (third millennium BC) Indus Valley, South Asia. The Indus (or Harappan) civilization of that era was urban, with cities like Harappa and Mohenjo-daro (e.g., Kenoyer 1991, 1998). These nucleated settlements were laid out on a grid plan with drain systems, elaborate craft industries, a largely undeciphered script, and monumental constructions of various kinds.

Drawing on the earlier works of Fairservis (1961, 1967) and Shaffer (1982), Possehl (1998) recently has questioned whether the ancient societies of the Indus were indeed states. Possehl begins his argument by developing criteria for states that are heavily influenced by the aforementioned theoretical ideas about centralization, as well as analogies with the Early Dynastic period in Mesopotamia and Pharaonic Egypt. Possehl (1998) consequently expects to find palaces, elaborate graves and burial monuments, marked stratification of wealth, elite aggrandizement, and massive statuary that personifies specific rulers with states. Yet, carefully weighing the evidence from the Indus, Possehl finds quite the opposite: no palaces, little evidence for individual aggrandizement or portraiture, no royal burial crypts, and no mention of specific rulers. He concludes that the Indus civilization was basically a "faceless culture," a complex society, but not a state.

Possehl (1998) suggests that political issues could have been resolved through decentralized councils and negotiation. Yet he also proposes that the defining quality of the Indus civilization was a distinct shared ideology that could have been used to give the society's diverse economic (and likely ethnic) segments a common ground and bind them together. Noting the highly repetitious nature of everyday objects in the residential areas of Indus cities, Miller (1985) has made a related argument for the importance of ritual in the creation of order in ancient Indus society. Rituals of fire and water evidenced on platforms in the great urban center may have been reproduced in everyday domestic rites. Masks were a prominent part of the cultural repertoire as they were at Teotihuacan.

I concur with Possehl (1998) when he recognizes the marked organizational differences between Indus society and its neighbors to the west. Yet I have no difficulty seeing the Indus centers, with their gridded streets and drainage systems, as capitals of states, albeit states with a more corporate form of organization along the lines of what he and Miller propose. This interpretation parallels what Stein (1994b) has proposed, at a lower level of hierarchical complexity, for the prestate Ubaid of late sixth–fifth millennium Mesopotamia. Stein (1994b:37–38) uses several lines of evidence (settlement pattern hierarchies with two levels, sites of 10 ha or more, the presence of temples, and a somewhat unequal distribution of certain artifacts) to argue that Ubaid, the immediate precursor to the development of states in Mesopotamia, was a complex, hierarchical society. Yet, in reviewing the data on Ubaid, Stein also poses the rhetorical question, "where's the chief?" There are no pronounced differences in elite-versus-commoner burials, no clear iconographic evidence for a chief, few high-status trade goods, and an absence of clearly elaborate residences of the sort we might associate with a chief (in the sense of Service). Instead, Stein (1994b:43) notes the remarkable stability and uniformity of Ubaid material culture, which he sees as reflecting a strong ideology. For Ubaid, Stein (1994b:35) proposes that the leaders of "small scale, irrigation-based Ubaid polities used inclusive ideologies emphasizing group membership through a strategy of ritually mobilized staple finance instead of overtly hierarchical principles."

Recently, both Mills (1997) and I (Feinman 2000) have independently advanced the corporate model as a new perspective on precontact Puebloan organization in the American Southwest (see Chapter 3, this volume), where acrimonious debate has long waged over the traditional notion of the egalitarian Pueblos. As in some of the previous cases discussed above, archaeologists have found little evidence for marked accumulations of wealth, princely burials, elite residences, or clear depictions of individualized rulers in the ancient pueblos. Yet, at the same time, certain ancient Puebloan societies were characterized by multitiered settlement systems with key differences between large and small sites, monumental constructions with elaborate ceremonial spaces, intensive agriculture, surplus agricultural production, and the uneven distribution of certain classes of labor-intensive artifacts. Based on these seemingly contradictory strands of evidence, archaeologists have pondered whether the precontact Pueblos were egalitarian or hierarchical. The concept of corporate hierarchies may provide a viable alternative (Feinman 2000) to the now polarized extant positions that emanate from the monolithic model. Several empirical observations seem to support this view, including the modular division of larger pueblos into component segments (e.g., Johnson 1989), the strong emphasis on community labor investments in ceremonial architecture, and indications that elements of social control may have been achieved through ritual (e.g., Adams 1991). Access to power appears to have been inherited, but often in a nonlinear fashion (e.g., Howell 1994).

Through the above cross-cultural comparisons, I have endeavored to illustrate that hierarchical formations can exist without a high degree of centralization, blatant expression of economic stratification, strong descent rhetoric, or highly personalized leadership. When these familiar aspects of more centralized network organization are absent, more corporate organizational modes seem to be found, which emphasize political offices rather than specific personages. In such cases, corporate codes that may help achieve order and integration through ritual are frequently present. Significantly, if we recognize the existence of corporate forms of organization, then we must broaden and reevaluate the correlates of hierarchical societies (like palaces, elaborate burials, marked disparities in wealth, and individualized rulers) that are so familiar to those of us who study sociopolitical behavior in the archaeological record.

CONCLUDING THOUGHTS

The corporate–network dimension introduced here is offered as a means to enhance and expand comparative evolutionary approaches rather than as a replacement for that school of thought. For comparative analyses, hierarchical complexity remains a key and distinct axis. Yet the corporate–network dimension provides a means to account more generally for specific variation within levels of sociopolitical complexity. As we have seen from the cases discussed above, corporate formations have tended to create interpretive enigmas for scholars reliant for explanation on the axis of complexity alone. The consideration of the orthogonal corporate–network dimension serves to broaden and strengthen our analytic arsenal and hopefully to build a better theoretical foundation for our study of societal variation and change.

I also wish to stress that corporate political formations are not envisioned here as utopian democracies or idealized communalistic societies. I cannot imagine how an urban society could operate effectively as a "committee of the whole." It is difficult enough for units as large and as "important" as executive committees of academic departments to "work" in this fashion. Certainly, wealth and power were not equally distributed among all inhabitants at Teotihuacan or Harappa or even the Ubaid and Puebloan settlements discussed above. Even within the corporate councils or power-sharing arrangements proposed here, certain individuals likely wielded more influence and clout. Yet if such happenings occurred, it would not override the recognized differences between corporate and network formations. In the former, certain sociocultural mechanisms clearly were in place to regulate and limit the degree of individual glorification, exclusionary politics, and personal accumulation that could be expressed. Such checks on personal display and aggrandizement are less in evidence in more network contexts. In corporate systems, the political world worked in markedly different ways than it did in network contexts where more exclusionary and individualizing forms of power were employed.

Admittedly, my principal comparison between Teotihuacan and the Classic Maya is intended to illustrate the poles of the corporate–network dimension; the framework certainly is not introduced as yet another inflexible typology. In this discussion, I have emphasized the continuous nature of the corporate–network dimension. Many "hybrid" cases with features of both strategies can be enumerated, including the late pre-Hispanic Aztec, a society that was in transition at the time of Spanish arrival. At that time, the political authorities of Aztec Tenochtitlán were not only consolidating their territorial domain, but a transition was in process from more network to more corporate forms of political action (Van Zantwijk 1985). As a consequence, the Aztec archaeological record has certain features that seem to parallel Teotihuacan (few ruler representations, relatively little variation in burial wealth, a strong cognitive code, and an urban architectural core at Tenochtitlán that focused on a large public space (see Blanton et al. 1996:11). At the same time, royal palaces were constructed (which have not been found at Teotihuacan), and Spanish inquisitors were able to construct a dynastic sequence. Yet power often was shared and the inheritance of political authority was anything but regularized or simple (e.g., Carrasco 1971; Paulinyi 1981). Aztec society was diverse, complex, and in transition. Any of these factors or others may account for the range of political–economic strategies that were employed.

The late Aztec example illustrates that the corporate–network dimension is a continuum (all spatiotemporal contexts do not cluster at one extreme or the other). Nevertheless, in this presentation I have emphasized the highly diverse organizational patterns of the Maya and Teotihuacan to establish the polar ends of the corporate–network dimension and to illustrate that such extremes existed contemporaneously in one particular global area (pre-Hispanic Mesoamerica). In doing so, I wish to avoid any simplistic tendency to affix rigid typological labels to specific peoples or cultures.

Rather, the corporate–network dimension represents strategies that could be implemented to different degrees over space and time, even within the same geographic setting or culture. For example, I have inferred above (see also Blanton et al. 1996) that early (pre-AD 200) Teotihuacan may have had more individualistic rulers and been somewhat more network in orientation. Likewise, most Preclassic and Postclassic Maya sociopolitical formations tended to be considerably more corporate in form than the Classic Maya.

We still have little understanding of what factors account for (or what conditions select for) corporate versus network modes. Even more significant for archaeologists, what circumstances prompt change in such organizational strategies? For example, what factors led to the rise and demise of the stela cult (and other aspects of network organization) that glorified Classic Maya rulers but was far rarer in the eastern lowlands before and after that era? Likewise, what set off the apparently radical restructuring of Teotihuacan (including the construction of hundreds of apartment compounds) that occurred sometime after AD 200 (Cowgill,1992b:107–109; Millon 1992:112)? In such "revolutionary episodes" (as with the transition from the Roman Republic to Empire), it seems reasonable to propose that the factional groups that challenged power wielders may have succeeded in overthrowing both the personnel and the fundamental sociocultural mechanisms that sustained their power. Yet such a hypothesis only serves to raise additional questions about the set of specific factors that may have been involved. Obviously, when it comes to understanding the cycling and development of leaders, rulers, and political centralization, many generations of concerted research are still required.

ACKNOWLEDGEMENTS

I would like to express a debt of gratitude to Jonathan Haas and the Field Museum for inviting me to participate in the stimulating conference that led to this volume and for their generous hospitality. I also would like to thank the conference participants for their thoughtful insights that led me to modify aspects of this chapter. Jonathan Haas, Richard Blanton, Linda Manzanilla, and Linda Nicholas read and commented on earlier drafts of this chapter for which I am deeply obliged. Glenn Storey provided me with invaluable assistance regarding ancient Rome, and he alerted me to the paper by Walters. Linda Nicholas assisted me on the production of the figures and tables as well as editorial matters. Although I deeply appreciate the assistance of all of these colleagues, any errors are entirely my own.

Chapter 9

Understanding the Timing and Tempo of the Evolution of Political Centralization on the Central Andean Coastline and Beyond

BRIAN R. BILLMAN

One of the fundamental questions in anthropology is the origin of formal institutions of rule. A central issue in the study of the origins of centralized political organization is the timing and tempo of the emergence of pristine chiefdoms and states. Why did centralized political institutions develop rapidly at an early date in certain areas, while in other areas they emerged more slowly at a much later date? Current views of human origins propose that anatomically modern human beings had evolved by approximately 100,000 years ago, yet centralized political organizations apparently did not develop for at least another 94,000 years. Early pristine chiefdoms and states

---

BRIAN R. BILLMAN • Department of Anthropology, University of North Carolina—Chapel Hill, Chapel Hill, North Carolina 27599.

From Leaders to Rulers, edited by Jonathan Haas. Kluwer Academic/Plenum Publishers, New York, 2001.

developed in only a few restricted areas between 6,000 and 3,000 years ago. Most of the world was not incorporated into state-level polities until the last 100 to 500 years during the period of colonial, nation-state expansion. Theories of the origins of centralized political organization must explain not only why centralized institutions developed, but how, when, and where. These crucial issues can be summarized in three questions:

1. How? By what process did pristine centralized polities develop?
2. Where? Why did this process occur in certain areas and not others?
3. When? Why did this process occur when it did in human prehistory?

For the purposes of discussion, in this chapter I pose three additional questions for investigating the process, timing, and spatial distribution of the formation of pristine polities:

1. Was the formation of pristine centralized polities a relatively simple process involving the interaction of just a few variables?
2. Was the process of political centralization influenced by a small set of human behavior characteristics?
3. Is it possible to define a relatively simple set of ecological and social conditions that governed the timing and geographic location of the emergence of pristine polities?

In short, is it possible to achieve a kind of unified field theory in anthropology that explains the process, timing, and spatial distribution of emergence of pristine centralized political organizations?

In order to investigate these questions, I use the central Andean coastline as a case study. During the Late Preceramic period (2700–1800 BC) and Initial period (1800–900 BC) some of the earliest large-scale monuments in the New World were constructed on the central Andean coastline (Burger 1992; Moseley 1975, 1992). Although these remarkable constructions long have been the subject of investigation, the nature of the political and social organizations that produced them remains one of the most controversial issues in Andean archaeology. Recent investigations at several of these early monumental centers strongly suggest significant variation in the degree of political centralization of the societies associated with these monuments, ranging from relatively acephalous, small-scale political organizations (Burger 1992) to large regional polities (S. Pozorski and T. Pozorski 1986, 1987, 1992). This implies substantial local and regional variation in the timing and tempo of the evolution of political centralization on the central Andean coast during the Late Preceramic and Initial periods. Variation is seen at the broad regional level and between neighboring valleys, such as the Casma and Santa Valleys. This variation in the timing and tempo of the formation of centralized polities may be the result of variation in ecological factors that affect the potential for economic control and surplus production. These factors include variation in (1) the regional and local impact of El Niño events, (2) the abundance of maritime resources, and (3) the extent of arable land in river valleys. By examining the relationship between these variables and the development of political centralization on the central Andean coast, the ecological and social conditions that promote political centralization can be investigated.

THE PROCESS OF CENTRALIZED POLITICAL FORMATION

Definition of terms is fundamental to understanding pristine political formation. Much of the debate over chiefdom and state formation has been at cross-purposes because researchers are discussing different processes or different aspects of the same process (Kristiansen 1991). Is the key issue the origins of social inequality, the formation of multiple levels of administrative hierarchy, or the development of surplus production and occupational specialization? Consider, for instance, the three most influential typologies of sociopolitical evolution: Service's (1962, 1975) band, tribe, chiefdom, and state; Fried's (1967) egalitarian, ranked, stratified, and state; and Polanyi's (1944) reciprocal, redistributive, and tributary. Each typology emphasizes different aspects of the same process: Service, social integration; Fried, control of resources; and Polanyi, exchange relationships. Further complicating the discussion, many have proposed that the use of typological schemes is entirely counterproductive (see, for instance, Earle 1978; Feinman and Neitzel 1984; Leonard and Jones 1987; Lightfoot 1984:2–5; Sebastian 1992:59–63).

Despite this sometimes all too apparent confusion over what we are talking about and what words we should use when we are talking about "it", since the emergence of the neoevolutionary school of thought over 50 years ago, consensus has begun to develop. Research has begun to focus on two interrelated processes: the development of centralized systems of political finance and political control (see, for instance, Billman 1999; D'Altroy and Earle 1985; Earle 1987a, 1994, 1997). All centralized political organizations require finance (the collection of goods or services to support political activities) and all centralized political organizations exercise some degree of control (the manipulation of the behavior of the general population through the use of political organizations). The fundamental challenge facing leaders of any emergent political organization is how to get households to contribute goods or services on a regular, predictable basis. Initially, aspiring leaders can use labor and resources from their own household to finance their political aspirations (Oliver 1970). However, if an emergent political faction is to grow, some means must be found to collect resources beyond the leader's household, a difficult task considering that small land-owning horticultural households are notoriously unwilling to produce beyond the needs of their members (Chayonov 1966; Pauketat 1996; Stanish 1999). Goods and labor are needed to finance political activities that are in turn used to maintain political control, that is, alter the behavior of an emerging political hierarchy and the general population.

This statement expresses the fundamental catch-22 facing aspiring leaders. A system of political finance is needed to exercise political control, and political control is needed to create a system of political finance. The situation is analogous to a young entrepreneur who finds that he needs to have cash or resources to get a bank loan to finance his project. Of course, if he had resources or cash, he would not need to ask for a loan.

The concept of political power is one means of understanding the circumstances under which leaders overcome the catch-22 of political finance and develop tributary relationships beyond their own households. Political power is the ability of one person—a leader—to get other people to do things (Adams 1975; Haas 1982; Mann 1986; Weber 1947). Political power relationships involve the interplay of at least nine variables: means, scope, amount, extension, power costs, compliance costs, refusal costs, gains, and base (Haas 1982).

In this chapter, I focus on just one of these nine variables: power bases. A power base consists of all those resources that a leader uses to change the behavior of other people (Haas 1982), for instance getting people to contribute goods or labor to a political organization. Types of political power bases are economic, ideological, and military. Economic power is based on the control over the procurement or distribution of basic resources or wealth goods. Direct physical control over the production or distribution of basic resources, such as food, clothing, or tools, can be achieved through the physical control of productive facilities, such as irrigation systems (Haas 1981, 1982; Moseley 1975), or by gaining a monopoly in trade in critical items, such as salt or high-quality stone for tool manufacture (Costin and Earle 1989; Haas 1981, 1982; Webb 1975). This differential access to basic resources provides political leaders with a source of economic power. By denying or rewarding households access to basic goods or productive facilities, leaders can create tributary relationships with households in order to finance political activities. This form of political control is also known as staple finance (Brumfiel and Earle 1987; D'Altroy and Earle 1985; Earle 1987a, 1994).

Economic power also can be achieved by obtaining a monopoly over the production or procurement of wealth goods, either by gaining a monopoly over foreign trade or by the direct control of local craft specialists (Brumfiel and Earle 1987; Earle 1978, 1987a, 1994; Gilman 1981; Haas 1981,

TIMING AND TEMPO OF POLITICAL CENTRALIZATION 181

1982). Differences in the productive capacity of households can be exploited by better-off households to gain access to wealth items. Households that are more productive, for instance, because they are larger or control more productive agricultural land, can generate surpluses that then can be used to acquire wealth goods. These wealth goods can be used to symbolize status differences and create networks of political allies and clients. Wealth goods are particularly usefully in creating and maintaining a hierarchical political organization. Providing access to such goods to lower-level leaders in an emerging political hierarchy is one means of ensuring continued allegiance of those leaders. Ultimately such a system of wealth finance can be used to further expand economic control and political centralization (Brumfiel and Earle 1987; D'Altroy and Earle 1985).

Although control of economic resources can yield considerable power to an aspiring leader, the circumstances under which economic control can be achieved are limited. In order to gain economic power, leaders must be able to intensify or monopolize the production or distribution of a resource. Because political organizations require surplus production for financing their activities, environments that permit sustainable increased production of a resource have the greatest potential for the formation of centralized political organizations. Sustainability of surplus production is crucial for the long-term stability of political organizations. For instance, nondomesticated food resources typically cannot yield strong economic power because increased harvesting of those resources often leads to a sharp decline in yields, such as when game is overexploited in an area. In contrast, in certain environments agricultural production can be increased without subsequent falloffs in yields through capital investments in production, such as irrigation, terracing, or ridgefields, or by the use of fertilizers or higher-yield crops. Although there are several well-documented cases of hunting and gathering groups developing chiefdom-level political organizations, these cases are limited to environments that have resources that can produce high sustainable yields. In the case of the Chumash on the central California (Arnold 1991; King 1990; Lambert and Walker 1991), the Calusa of Florida (Widmer 1988), and the Tlingit in southeast Alaska (de Laguna 1983; Drucker 1955; Maschner 1991), abundant maritime resources allowed leaders to extract large surpluses and develop some degree of economic power and political control.

More complex and larger political organizations develop when resources are available that are both intensifiable and monopolizable. For instance, the shift to irrigation agriculture allows both increased production and control because leaders potentially could monopolize access to irrigated land and water. In contrast, the degree of economic control and concomitant economic power exercised by leaders in Chumash, Calusa, and Tlingit societies was limited because marine resources are difficult to monopolize. In other words, it is difficult for leaders to deny people access to fish in the sea.

Ideological power is based on control over systems of belief. The most fundamental form of ideological power is charisma: the ability of a leader to use words and ideas to inspire or persuade followers to do what the leader wishes. Those individuals with the ability to move the masses through words can either manipulate existing belief systems or create new belief systems to persuade households to contribute labor or goods to an emergent political faction or organization. Two factors, however, limit the effectiveness of charisma for extracting surplus from households. First, prior to the development of mass media technologies, such as printing presses, radio, or television, the number of people that could be reached by the persuasive powers of an emerging leader was limited. Second, not only must the potential follower be able to physically receive the message, but also he must be receptive to it and volunteer to contribute. If ideological power is to be used to finance a political organization, some means must be found to extend the scope of the dissemination of ideological messages and to regularize the contribution of goods or labor by followers.

The number of people receiving an ideological message can be increased by the materialization of ideology (DeMarris et al. 1996). As defined by DeMarris et al. (1996:16), "Materialization is the transformation of ideas, values, stories, myths, and the like, into physical reality—a ceremonial event, symbolic object, or a writing system." Public ceremonies are one means of drawing large numbers of people together in one place, thus creating opportunities for charismatic leaders to influence a large segment of the local or regional population. Public monuments are another means of conveying messages to large groups. Monuments can be used to commemorate individuals or events or symbolize concepts that are part of the ideology of a political organization or faction. For instance, the Tomb of the Unknown Soldier and the depiction at US Marine Corps Memorial of the Marines raising the flag at Iwo Jima are potent symbols of self-sacrifice and military valor in the service of the state. Ideology also can be disseminated by objects with symbolic meaning (such as ritual paraphernalia or holy relics) and by writing or iconographic systems. Because written documents and symbolic objects can be portable, they allow political ideologies to be disseminated efficiently over large areas.

Ceremonial events, public monuments, symbolic objects, and writing or iconographic systems all are media for the expression of ideas beyond small local groups, and thus can be used by leaders to create shared belief systems that legitimize both the elevated status of the leaders and the necessity of regular, fixed contributions by followers. Because these media have a material form, in certain situations aspiring leaders can control or monopolize media of expression and create a potent ideological power base. The creation of formal ideological institutions, such as churches, pilgrimage sites, and monuments, allows leaders to monopolize sacred knowledge and formalize the dissemination of ideas and collection of tribute. When belief systems have become institutionalized, control over the ideological infrastructure potentially can allow leaders, even those who are "charisma challenged," to exercise formal positive and negative ideological sanctions and extract regular tribute payments from households. Positive sanctions can include granting access to sacred places, icons, rituals, or knowledge, whereas negative sanction can include excommunication or shunning.

Times of environmental, economic, and social crisis create potential for the manipulation of ideology for political ends. In such circumstances, people are more receptive to new ideas. When good times follow periods of stress, leaders are able to further solidify ideological control by claiming credit, whether deserved or not, for improved conditions. Other opportunities for the successful use of ideological power are created when exploited, oppressed, or less-well-off groups exist within or on the periphery of a polity. As we have witnessed in this century, liberation ideologies in the hands of a charismatic leader have the power to inspire groups to achieve extraordinary feats.

Military power is based on the control of military or police forces. Creation of a military power base is dependent on the recruitment, maintenance, and successful use of military or police forces. Although a military power base can be used to extract tribute from households by threats of violence, terrorizing and repressing members of one's own political organization are not often an effective political strategy. This is especially true in the early stages of development of an organization or faction, when the organization may be too weak to resist revolt or defection by members. A successful external military campaign is one way of using military power to develop tributary relations without terrorizing follower or potential follower households. During periods of external conflict, looted goods and captives from households outside an emergent political organization can be used to finance political activities. As a result, some or all of the cost of the political economy is externalized. In addition, in such situations leaders do not have to use the threat of violence to extract tribute from local households to finance their activities. Faced with aggression and violence from outside groups, households are more willing to contribute to leaders. This allows internal leaders to play "good cop" to the external "bad cop," while achieving their own political ends, that is, the extraction of surplus from local households to finance the leader's political aspirations. Because the use of military forces is a high-risk strategy, military power bases have several serious drawbacks that limit their use. In order to exercise military power, political leaders must find a way to get households to contribute surplus labor, in the form of warriors, for a highly dangerous activity with uncertain rewards. Virtually any armed conflict will involve the loss of life or serious injury of supporters. High or even moderate or low causalities may result in popular discontent or defection of allies or subordinate military commanders. In addition, the outcome of any battle or war is never certain. Smaller forces have been known to defeat large, technologically superior forces, as the United States and the Soviet Union discovered in Vietnam and Afghanistan. Defeat on the battlefield, defection of supporters, or popular revolt can result in the rapid demise of the political career of a military leader or even the collapse of a polity. Finally, the use of military force is often expensive, requiring the production of weapons, construction of fortifications, and provisioning of troops. Leaders must find some means of financing their military power base without overburdening their supporters. In sum, if raiding or warfare is to become the main power base of a political organization or faction, some means must be found to recruit raiders, achieve military victories, and extract tribute on a regular, predictable basis.

Although warfare is a risky business, in certain circumstances the probability of military victory by a particular group can be greatly increased. Military success can be achieved through innovations in logistics, organization, or weaponry; through superior leadership, strategy, and tactics; or through massing greater numbers of warriors. Although the fog of battle obscures the ultimate outcome of any conflict, larger groups with better weapons, organization, and logistics that employ superior leadership, strategy, and tactics are more likely to succeed in the end. Thus in situations where one group has all or some of those advantages, the risks of military conflict are reduced and opportunities for the use of military power by aspiring leaders are created.

The second problem, the regular recruitment of warriors, can be reduced in certain circumstances. The emergence of a charismatic leader can result in the recruitment of a large force. However, because charisma is not apparently an inherited trait, it is often limited to one generation of leadership. Times of conflict, whether caused by attacks by foreign groups, economic or ecological crises, or the deliberate acts of leaders, create opportunities for development of a military power base. In circumstances of real or imagined crisis, aspiring leaders are more likely to successfully recruit warriors for military actions. For instance, in the face of raiding by outside groups, households are more likely to turn to potential military leaders for help, contributing men and goods for the defense of the community. Times of conflict also can cause populations to aggregate for self-defense. Aggregation

TIMING AND TEMPO OF POLITICAL CENTRALIZATION

creates further opportunities for leaders to mass larger military forces (Chagnon 1983; Earle et al. 1987; Haas and Creamer 1993). Use of larger forces, in turn, can force further aggregation by opponents. Each round in this escalating feedback cycle further strengthens the military power base of leaders.

If leaders are able to amass and use a military force successfully, then several means exist for regularizing the collection of goods and labor needed to finance political organizations. In the case of the Wolof State in west Africa, war captives were put to work on state lands in order to produce agricultural goods to finance state activities (Tymowski 1991). In other cases, adjacent regions can be intimidated into tributary relationships. Finally, direct conquest and control also can be used to extract tribute from nearby territories (D'Altroy 1994; Carneiro 1970, 1972, 1981).

Political finance is ultimately rooted in the economic, ideological, and military power bases controlled by leaders. Control of power bases provides leaders the means by which they can manipulate followers into contributing goods or labor to finance a political organization. By tapping into particular power bases, aspiring leaders devise strategies for creating or gaining control of existing political organizations. In actual practice, no political strategy relies purely on one power base because all three types of power bases are interrelated. For instance, the successful conquest of neighboring regions through the use of military power can greatly enhance the economic power base of leaders through the confiscation of arable land or other resources in conquered areas. Conversely, surpluses created by an economic power base, such as the control of arable land or trade, can be used to raise an army or hire mercenaries, thus expanding the military power base of leaders. Economic resources also can be used to increase ideological power by financing public rituals, feasts, or monuments that legitimize authority and disseminate political propaganda.

Arguably, ideology, particularly in the form of charismatic leadership, is at the start of all political organizations regardless of the power bases ultimately controlled by emerging leaders. In the initial stages of political formation charismatic leadership can inspire men to go to war or families to contribute economic resources. Although ideology may play a dominant role in the early stages of political formation, ultimately the particular strategy used by a political leader is dependent on the opportunities for control of economic, military, or ideological resources. Therefore, the key to understanding the timing and tempo of emergence of pristine centralized political organizations on the central Andean coastline—and elsewhere in the world—is determining the specific social, economic, and ecological circumstances that permit leaders to gain control over economic, military, or ideological power bases.

THE CENTRAL ANDEAN COASTLINE

The coastline of the central Andes has one of the driest deserts, richest sea fisheries, and highest mountain ranges in the world. This unique juxtaposition of narrow coastal plain, high mountains, and cold water upwelling of the Humboldt Current creates one of the most productive agricultural and maritime environments in the world. The Humboldt Current is laden with organic nutrients and plankton, which support an abundance of marine life. Approximately one-fifth of the world's commercial fish harvest comes from the central Andean coastline (Moseley 1975). Moisture collected offshore by the southeasterly winds falls on the mountains and returns to the coastal plain as runoff in permanent and seasonal rivers, creating river oases in the coastal desert. Water from these rivers has nourished agricultural fields on the coast for thousands of years. While the apparent marine and terrestrial abundance of the central Andean coast created opportunities, environmental perturbation also placed critical constraints on indigenous populations. The volume of river flow varies dramatically from year to year, and seasonal variation in volume limits the extent of double cropping. In addition, several times a century severe climatic reversals, known as El Niño events, occur on the coast. During these events, the Humboldt Current diminishes in strength and the warm water El Niño countercurrent surges southward, causing massive die-offs of cold water fish and rain and flooding on the coast, devastating both agricultural and marine production (Quin and Neal 1987; Waylen and Caviedes 1986). La Niñas, which are extreme cold ocean temperatures, also occur periodically, causing droughts that limit the extent of irrigation on the coast (Waylen and Caviedes 1986).

For the purposes of this chapter, the central Andean coastline is defined as the area from the Tumbes Valley at the modern Peru–Ecuador border to the Moquegua River, near the Peru–Chile border (Fig. 1). The northern limits correspond to the transition from dry coastal desert to mangrove swamp. The southern limit is more arbitrarily set at a point roughly parallel to the Titicaca Basin. In this chapter, I subdivide the central Andean coast into north, central, and south sections (Fig. 1). Because the central Andean coastline roughly corresponds to the coast of Peru, these sections will be referred to as the north, central, and south coast of Peru. Although the environment is generally similar across the length of the central Andean coast, two important north–south environmental trends are present. First, the frequency of El Niño events decreases as one moves from north to south. The central coast of Peru experiences El Niño events less often than the north coast, and the south coast rarely suffers the effects of these events. Events severe enough to affect the central coast occur every 14 to 64 years, whereas moderate to strong events occur along the north coast of Peru every 4 to 10 years (Parsons 1970; Quin and Neal 1987).

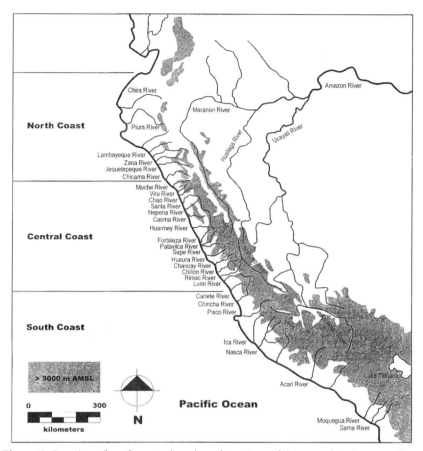

Figure 1. Location of north, central, and south sections of the central Andean coastline.

A second environment trend concerns the quantity of arable land in river valleys. The extent of arable land is largely determined by the width of the coastal plain. Moving from north to south the Andes is located closer to the sea. As a result the coastal plain narrows. On the north coast of Peru the coastal plain is over 100 km wide, whereas on the south coast the coastal plain disappears entirely. Consequently, north coast river valleys have upward of 100,000 ha of arable land, and central coast valleys range from a few thousand to 20,000 ha. South coast valleys have a few thousand hectares of land located in pockets many kilometers inland.

The result of these two trends is that while the north coast of Peru has vast tracts of arable land, marine resources are subject to frequent perturbations. In contrast, the south coast has rich, stable marine resources, but arable land is limited to small pockets located well inland. The central coast appears to represent a happy median between these trends. As we will see, the central coast of Peru was the portion of the central Andean coastline that was most amenable to the formation of early, centralized polities. The crucial questions are why did the earliest centralized polities emerge along the central coast of Peru? What set of specific social and ecological conditions permitted leaders to gain control over economic, military, or ideological power bases and form centralized polities at such an early date?

THE LATE PRECERAMIC PERIOD

Several important changes in subsistence, settlement, and sociopolitical organization occurred during the Late Preceramic period (2700–1800 BC) on the central Andean coastline (Burger 1992:27–56; Moseley 1975). Changes in subsistence and settlement include the introduction of domestics, intensification of fishing, and a shift to sedentary village life. Changes in social and political organization are indicated by the presence of small- and moderate-sized ceremonial structures, including mounds, terraced hillslope complexes, and circular sunken courts.

Although the period preceding the Late Preceramic period is poorly understood, the start of the Late Preceramic period was apparently marked by increased exploitation of marine resources and the introduction of several cultigens. At Padre Alban, a site dating to the early part of the Late Preceramic period in the Moche Valley, mollusks comprise more than half the animal protein represented in the archaeological samples, followed by fish (27%), marine birds (20%), and sea lions (*Otaria byronia*) (S. Pozorski 1976, 1983). Only three cultigens were found at the site: squash (*Cucurbita* sp.), gourds (*Lagenaria siceraria*), and cotton (*Gossypium baradense*). Because only one food crop was present, Shelia Pozorski proposed that cultivation was primarily oriented toward the production of nonfood crops. Gourds were grown for containers and net floats and cotton was grown for netting. Further, she proposed that the subsistence base of the inhabitants of the site was transitional between a terrestrial-based hunting and gathering economy and a farming and fishing economy.

By the latter part of the Late Preceramic period, the subsistence system is characterized by increased exploitation of fish and cultigens. In contrast to Padre Alban, at Alto Salaverry, which also is located in the Moche Valley, 42% of the animal protein represented in midden deposits was derived from fish, while mollusks comprised 38%, sea lions 13%, and marine birds 7% (S. Pozorski and T. Pozorski 1979). Mollusks at Alto Salaverry generally are much smaller and younger than those found at Padre Alban, suggesting overexploitation of local shellfish beds. At both sites, the overwhelming majority of the fish exploited were near-shore rather than offshore species. Most striking is the increased use of cultigens. In addition to cotton, gourds, and squash, which are abundant at the site, seven new species were found (S. Pozorski and T. Pozorski 1979). These are, in decreasing order of occurrence, cansaboca (*Bunchosia armeniaca*), common beans (*Phaseolus vulgaris*), lucuma (Lucuma obovata), avocado (*Persia americana*), pepper (*Capsicum* sp.), pacae (*Inga feuillei*), and guava (*Psidium guajava*).

The Late Preceramic period also is characterized by a shift to sedentary villages. Numerous large coastal Late Preceramic period villages with thick midden deposits and, in many cases, public architecture have been documented on the central Andean coastline (Fig. 2). The distribution of large

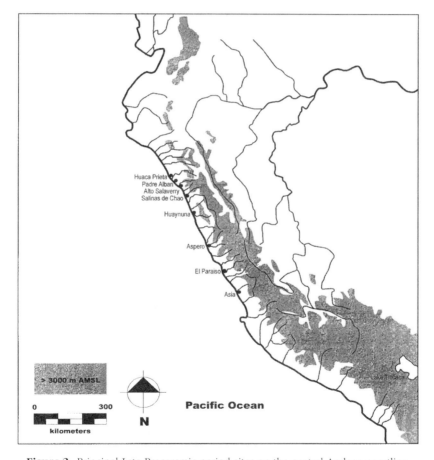

Figure 2. Principal Late Preceramic period sites on the central Andean coastline.

sedentary villages, however, was not uniform along the coast. Three patterns are apparent in the regional distribution of large villages. First, virtually every river valley on the central coast of Peru from the Chicama River to the Asia River had one or more large Late Preceramic period villages. North of the Chicama Valley and south of the Asia Valley, large sedentary villages were less common. Second, the size of the largest settlement in each valley generally increases as one moves south. Village size peaks in the Chillón Valley just north of Lima and then declines to the south (Table 1). Investment in public architecture follows a similar pattern, with the largest Late Preceramic period monumental construction, El Paraíso, located in the Chillón Valley (Moseley 1975; Quilter 1985) (Table 1). No Late Preceramic period public architecture has been found on the coast north of the Moche Valley or south of the Chillón Valley. The layout and size of coastal Late Preceramic public architecture varies greatly from valley to valley. The size of public architecture ranges from the small circular sunken court at Alto Salaverry, which is only 9 m in diameter and required only 115 m^3 of excavation (S. Pozorski and T. Pozorski 1979), to the massive mounds at El Paraíso, which has an estimated volume of 340,000 m^3 (Burger 1992:28; Moseley 1975; Quilter 1985). Types of public architecture include U-shaped mounds, terraced structures, and sunken courts. The pattern of monumental construction in the Late Preceramic period suggests that leaders in the central coast valleys of Peru from the Moche to the Chillón Valley controlled much more labor than north or south coast leaders.

Based on the layout, these ceremonial structures appear to have been used primarily for public displays (Moore 1996a,b). Mounds and terraced structures appear to have served as platforms for the display of public ritu-

Table 1. Summary of Published Data on Sites Dating to the Latter Half of Late Preceramic Period on the Central Andean Coastline

Valley	Site	Size (ha)	Depth (m)	Public architecture	Reference
Chicama	Huaca Prieta	>1.0	>1.0	Absent	Bird (1948)
Moche	Alto Salaverry	1.11	1.5	Sunken court	S. Pozorski and T. Pozorski (1979)
Viru	Huaca Prieta de Guañape	3.3	>1.0	Absent	Willey (1953)
Chao	Salinas de Chao	8.0	?	Terraced complex	Alva (1986)
Santa	SVP-SAL-7	8.75	Unknown	Absent	Wilson (1988)
Casma	Huaynuna	8.5	2.0	Terrace complex	T. Pozorski and S. Pozorski (1990)
Supe	Áspero	13.19	2.0	6 mounds	Feldman (1987)
Chillón	El Paraiso	58	Unknown	9–13 mounds	Moseley (1975); Quilter (1985)
Asia	Asia	?	?	Absent	Engel (1963)

als. Based on their design, sunken courts also could have been used as meeting places for the presentation of rituals. If these monuments were designed for public displays, the audience capacity of Late Preceramic monuments increased dramatically from north to south. The sunken court at Alto Salaverry had a seating capacity of approximately 57 to 85 people (Billman 1996:113). An audience of several hundred could have viewed rituals at the terraced structure at Huaynuná in the Casma Valley (T. Pozorski and S. Pozorski 1990) and the mounds at Áspero in the Supe Valley (Feldman 1987). At El Paraíso in the Chillón Valley, an audience numbering in the thousands was possible (Moseley 1975; Quilter 1985).

In addition to public displays, monuments such as Áspero and El Paraíso also contain smaller rooms with restricted access. These rooms may have been used for nonpublic rituals or other nonpublic activities. Feldman (1987) has suggested that the development of nonpublic spaces within public architecture may indicate the emergence of an elite priestly class.

In sum, two important changes occurred during the Late Preceramic period on the coast: the formation of large sedentary villages and the construction of the first public monuments in the Andes. Alto Salaverry in the Moche Valley may be the northern-most, large sedentary village on the coast, and Asia, just south of the Lurin Valley, may have been the southern-most village. The emergence of both phenomena (sedentary villages and monumental construction) may correlate with the emergence of the first centralized political organizations on the coast.

THE INITIAL PERIOD

The Initial period (1800–900 BC) was a period of dramatic technological, demographic, and sociopolitical change. During this period ceramic production and irrigation were introduced, the number of sites increased exponentially, and numerous large-scale public monuments were constructed.

A fairly detailed picture of the Late Preceramic–Initial period transition is available from the Moche Valley. A pedestrian survey has been completed of the valley (Billman 1996), and excavations have been conducted at two Initial period sites: Caballo Muerto (the largest early ceremonial center in the valley), and Gramalote (a coastal village) (S. Pozorski and T. Pozorski 1983; T. Pozorski 1976, 1980, 1982). The Initial period in the Moche Valley can be divided into two phases (Early and Middle Guañape), which date to 1800–1300 BC and 1300–800 BC, respectively. The Late Guañape Phase corresponds to the Early Horizon and dates to 800–400 BC

Settlement pattern data from the Moche Valley indicate a sharp increase in number of habitation sites, a shift in settlement from the coast to the middle valley, and increased reliance on irrigation agriculture. Between the latter half of Late Preceramic period and the Late Guañape phase the total area of known habitation sites increased from 0.5 to 15.81 ha, an increase of over 3100% during a 1800-year period (Billman 1996:202–205). Changes in the distribution of sites were equally dramatic. All three Late Preceramic sites were located on or near the coast. In the Early Guañape phase the number of known sites increased to 24, but only one site was located on the coast (Gramalote). The other 23 sites were located in the middle valley. In the Middle and Late Guañape phases no sites were located within 10 km of the coast. Most of the sites in the Guañape phases were located in the middle valley between the valley neck and the confluence of the Moche and Sinsicap Rivers. This shift in settlement to the middle valley may have been caused by the introduction of irrigation. With the introduction of irrigation agriculture, the middle valley was preferred because it can be irrigated with relatively short canals. In contrast, the alluvial fan of lower valley can be irrigated only through the use of long canals (Billman 1996:164–167).

Investment in public architecture also dramatically increased in the Moche Valley from the Late Preceramic period to the Early Guañape phase. Only one Late Preceramic period public construction, the circular sunken court at Alto Salaverry, is known in the valley (Billman 1996:113; S. Pozorski and T Pozorski 1979). The volume excavated to construct the sunken court was only 115 m^3. In the Early Guañape phase, one large mound complex, one terrace and mound complex, two small isolated mounds, and two small terraced structures were constructed for a total of 33,590 m^3 of public architecture, an increase of 292 times (Billman 1996:168–179, 1999). In the Middle Guañape phase, the number of ceremonial sites and the volume of construction increased dramatically again. In this phase, one large, two medium, and three small ceremonial centers were constructed for a total of between 283,003 and 318,667 m^3 of public architecture (Billman 1996:168–179, 1998). This is eight or nine times the volume of construction during the Early Guañape phase.

Although the introduction of irrigation agriculture, shifts to middle valley settlement, and dramatic demographic expansion occurred across most of the central Andean coastline during the Initial period, not all river valleys experienced an exponential increase in monumental constructions (Fig. 3).

North of the Moche Valley, few Initial period monumental constructions have been identified, and those few that have been recorded are small in scale, generally ranging from less than 1,000 m^3 to 10,000 m^3. For example, in the Chicama Valley, which is the next valley north of the Moche Valley, only five small Initial period mounds have been identified in the valley (Leonard 1995; Leonard and Russell 1994). Each mound encompasses

TIMING AND TEMPO OF POLITICAL CENTRALIZATION

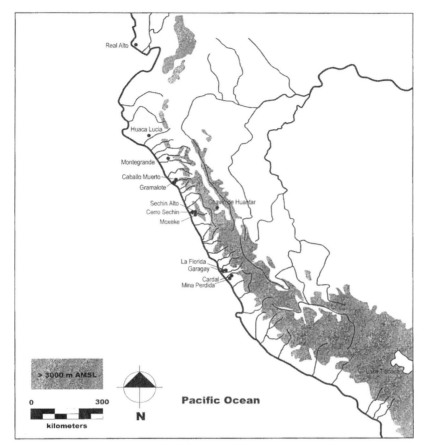

Figure 3. Principal Initial period sites on the central Andean coastline.

less than 10,000 m³ of fill. In the Jequetepeque Valley, the next valley to the north, no large or moderately sized early ceremonial centers have been identified. Surveys have documented 30 early ceramic sites in the middle Jequetepeque Valley that contain pubic architecture (Ravines 1985). Those monuments range from single platforms to complexes of terraces and platforms, and all are less 10,000 m³ in volume. North of Jequetepeque, the only Initial period monument that has been recorded is Huaca Lucia, a small mound in the Lambayeque Valley (Shimada et al. 1982).

On the central coast of Peru between the Moche and Lurin Valleys, several valleys contain monuments that far exceed the Initial period monumental construction on the north coast of Peru (Burger 1992; Moseley 1975). In the Casma Valley, the early Initial period site of Pampa de Las Llamas–

Moxeke contains two primary mounds, Moxeke and Huaca A, that measure 170 by 160 by 30 m tall and 125 by 119 by 12 m tall, respectively (S. Pozorski and T. Pozorski 1986). In addition, the site has 107 smaller U-shaped mounds the range from 20 to 50 m on a side and 2 to 5 m tall. The volume of Moxeke, which is estimated at 500,000 m^3, alone exceeds the total volume of Initial period monumental construction in the Moche Valley. The small U-shaped mounds are similar in size to the Initial period monuments north of the Moche Valley. The volume of all of the 107 U-shaped mounds at the site may well exceed the total volume of Initial period monumental construction on the entire coast of Peru north of the Moche Valley. Pampa de Las Llamas–Moxeke was apparently replaced by Sechin Alto, an even larger center, in the latter part of the Initial period. Located on a tributary of the Casma River, the main mound at Sechin Alto measures 200 by 250 by 34 m tall, and lesser mounds and plazas extend fully 1.4 km out from the main mound (S. Pozorski and T. Pozorski 1987). The total volume of construction at Sechin Alto is over 2,000,000 m^3. Altogether, several million cubic meters of monumental construction occurred in Casma Valley during the Initial period.

Further south on the central coast several other massive monuments were constructed in the Initial period. In the Chillón and Rimac Valleys, Huaca la Florida was constructed in the early Initial period (Patterson 1985). The Huaca is 17 m tall and has two 500-m long and 3- to 4-m tall wings. Total fill of this structure is over 1,000,000 m^3. Patterson (1985) estimates that 500 to 1000 people constructed this structure in 200 years. The construction of Huaca la Florida was followed by La Garagay, which has close to 400,000 m^3 of fill (Patterson 1985).

This pattern of massive construction in the Initial period also is apparent in other less well-studied central coast valleys. Numerous large Initial period sites have been documented in the Nepeña, Fortaleza, Pativilca, Supe, Hauara, Chancay, and Lurin Valleys (Burger 1992:56–103).

On the south coast of Peru, the pattern of monumental construction is remarkably different. South of the Lurin Valley, not a single Initial period monumental construction has been identified, despite the completion of several large survey projects in the Ica, Nasca, and Moquegua Valleys (Owen 1993; Schreiber 1999; Silverman 1993, 1994). In the Ica Valley, surveys have identified aggregated sites and site clusters that apparently were the manifestation of a series of small autonomous chiefdoms (Massey 1986). Recent surveys in the Nasca Valley have for the first time identified Initial period and Early Horizon sites in the valley; however, no public architecture has yet been found (Schreiber 1999; Silverman 1993, 1994). Monumental construction apparently did not occur in the Nasca Valley until the founding of Cahuachi in the Early Intermediate period.

Crucial to understanding Initial period political organization on the central Andean Coast is understanding the function of public architecture in this period. Initial period monuments were more than just labor projects distributed on the landscape. They served some societal function. Excavations at Initial period monuments in the Moche and Casma Valleys indicate that these monuments were not used as residences or tombs (S. Pozorski and T. Pozorski 1986; T. Pozorski 1976, 1980, 1982). The design of these structures, which is remarkably consistent from valley to valley, suggests they may have been used primarily for public displays or rituals. The mounds typically have an entrance that leads up to one or more platforms, and plazas or sunken courts are located in front of the platforms. At Huaca de los Reyes and Moxeke, the platforms had large elaborate friezes (S. Pozorski and T. Pozorski 1986; T. Pozorski 1976, 1980, 1982; Tello 1956). In essence, these monuments were elevated platforms fronted by large public spaces and would have been ideally suited for presentations of public displays and rituals (Moore 1996a,b).

Although virtually all Initial period ceremonial centers on the coast have large public spaces, many of the larger mounds have enclosed spaces on top that were beyond public view. At Huaca de los Reyes, at the site of Caballo Muerto in the Moche Valley, Thomas Pozorski (1976, 1980, 1982) noted a three-tier hierarchy of space: (1) two large open plazas located in front of and between the wings of the structure, (2) a smaller plaza above these plazas, and (3) a sunken gallery at the highest level of the structure. Access to these spaces was progressively more restricted and audience capacity progressively smaller. Thomas Pozorski (1976, 1980, 1982) proposed that this hierarchy of space mimicked a three-tiered social hierarchy with access to ritual knowledge restricted to the highest levels of the social hierarchy.

LEADER–FOLLOWER RELATIONSHIPS IN THE LATE PRECERAMIC AND INITIAL PERIODS

Although changes in subsistence permitted the formation of large sedentary villages in the Late Preceramic period, changes in subsistence alone do not explain why such villages came into existence. Large social groups were not required to exploit floodplain agriculture or near-shore marine environments. Individual households or small household clusters could have flourished with such a subsistence system without the increases in disease and social tension that are associated with village life. Fishing and shellfish collecting could have been conducted effectively by individuals or small groups (S. Pozorski and T. Pozorski 1979). For instance, near-shore fishing in the Moche Valley currently is done from small one-person boats (caballitos de totora). Floodplain agriculture involved slashing and burning small plots and could have been carried out by individual families (Moseley 1975; Parsons 1970).

One explanation for village formation is warfare, which can force households together out of the need for mutual defense. However, there is little evidence of armed conflict in the Late Preceramic period on the coast. There is no apparent preference for defense settings for site locations or evidence of fortifications. In addition, burial populations show little evidence of violent conflict, although the sample size currently is small (Quilter 1985; Vradenburg et al. 1997). The only direct evidence of early conflict comes from the coastal site of Ostra in the Santa Valley (Topic 1989; Topic and Topic 1987). This site predates the Late Preceramic period, ca. 3000 BC, and has two rows of possible slingstone piles stretching for 75 to 100 m. Even if this is a case of early conflict, the site stands alone as the only known example of preceramic warfare on the central Andean coast.

An alternative explanation for the formation of sedentary villages in this period can be found in the realm of politics. One important aspect of the near-shore environment is its high productivity. The abundance of fish is such that Late Preceramic period populations utilizing simple techniques, such as line fishing, haul nets, or gill nets, could not have overfished the coastline. Today, despite 5000 years of use and decades of fishing with modern techniques, the near-shore marine environment of the coast of Peru still sustains high fishing yields. Consequently, individuals in the Late Preceramic period could have intensified harvesting of fish and created large surpluses without causing a sharp decline in overall production. Surpluses could have been used by households as a buffer against El Niño or other low productivity years, either by storing the goods directly or by creating reciprocal exchange relationships.

In the hands of aspiring leaders, surplus production could have been used to increase their status and influence. Surpluses could have been used as a fund of power that potentially could have been spent in a number of ways. For instance, with these funds leaders could have financed dances, feasts, ceremonies, or public monuments; been invested in regional trade to obtain status goods; or created obligatory relationships between donor and recipients. As evidenced by the numerous ceremonial structures on the coast, construction of public monuments and the staging of public rituals appear to have been primary uses of the surpluses controlled by leaders.

A critical challenge for Late Preceramic period leaders on the coast would have been maintaining a coalition of followers. The absence of intensive armed conflict provided few opportunities for leaders to exercise military power and the productive system during Late Preceramic period provided few opportunities for economic control. Agricultural land was still abundant and required slash–burn rather than irrigation techniques. Simple technologies provided easy access to near-shore resources for any fit individual. Consequently, stores of dried fish amassed by leaders would have been of limited use for controlling followers. Because families had easy access to fish and shellfish, they would have had little incentive to do as leaders urged, if their only reward was simply more dried fish. The exception to this would have been during El Niño, when increased ocean temperatures devastate cold-water fish and shellfish populations. During such environmental disasters, ill-prepared families may have been forced to turn to leaders who had large stores of food in order to survive, thus creating debt obligations. These environmental disasters also created opportunities for leaders to exercise ideological power. As noted previously, times of environmental crisis create potential for the manipulation of ideology for political ends. When good times followed periods of El Niño disruption, leaders may have been able to further solidify ideological control by claiming credit, whether deserved or not, for improved conditions.

In sum, during the Late Preceramic period, leaders probably had to rely on ideological power (especially personal charisma) and redistribution of surpluses produced by their own households in order to create and maintain leader–follower relationships. Public architecture and large storage facilities at Alto Salaverry (Billman 1996:104–112), and other Late Preceramic period sites, may be a manifestation of status striving and coalition building by early leaders. In this scenario, early villages and public architecture on the central Andean coast were the result of political activities by aspiring elites.

The exponential increase in the construction of monuments in the Initial period strongly suggests that leader–follower power relationships changed dramatically in certain river valleys (Haas 1987). The huge increase in volume of construction indicates that certain leaders in the Initial period were able to mobilize much larger labor forces than leaders in the Late Preceramic period. Changes in leader–follower relationships may have been linked to the expansion of two power bases controlled by Initial period leaders. In the Initial period political power may have been based on (1) the control of irrigation canals, water, land, and the surpluses produced by irrigation, and (2) the control of ceremonial monuments, ritual knowledge, and ritual activity.

With the shift to irrigation in the Initial period, emerging elite households had new opportunities for the accumulation of surplus labor and economic control (Haas 1987; Moseley 1975). Although the initial cost of constructing irrigation systems was high relative to fishing and floodplain horticulture, once in production irrigation systems dramatically increased yields. By organizing and financing the construction of irrigation systems, emerging elite households could have reaped high yields once the systems were in production. Surplus production extracted from those irrigation systems could have been used to fund further political activity and to create economic differences between households. Irrigation systems also created opportunities for individuals or families to control land and the flow of life-giving water, another possible source of political and economic control. The consequence of this was that in the Initial period political leaders not only controlled increased labor and goods, they also controlled the distribution of land and water (Haas 1987; Moseley 1975).

With this newfound source of surplus production and political control, leaders were able to finance the construction of large public monuments at which public and private rituals were conducted. Investments in public architecture point to the importance of ideological power for Initial period leaders. In the Moche Valley, for instance, the volume of monumental construction far exceeded the volume of canal construction by perhaps as much as nine times (Billman 1996:344–345, 1999). For Initial period leaders, public monuments and rituals undoubtedly served to legitimize their authority and the extraction of surpluses and enabled them to disseminate their political ideology to local and regional populations. Public monuments and rituals materialized ideology in a form that could be controlled by leaders. By controlling access to these monuments and the content of rituals conducted at them, leaders created a strong ideological power base. Denying and rewarding people with access to public and private rituals provided potential positive and negative sanctions for controlling behavior, especially the behavior of political allies inside and outside of the polity. These sanctions would have been particularly useful for maintaining the allegiance of lower level administrators, which is essential for creating and controlling a multilevel administrative hierarchy. Further, control of the content of public rituals provided a means of manipulating the behavior of the general population.

The pristine Initial period polities on the coast may have been the result of a feedback relationship between economic and ideological power. The surplus production and increased political control afforded by the expansion of irrigation allowed increased investment in the ideological infrastructure, which in turn legitimized further extraction of more surplus goods and labor. Each incremental increase in economic power allowed an incremental increase in ideological power that in turn allowed a further increase in economic power, and so on, until leaders controlled the labor of thousands of people and directed the construction of the some of the largest monuments in the New World.

A possible weakness of these Initial period leaders was their limited control of a third power base: military power. There is little evidence of warfare on the coast in the Initial period. Although murals at Cerro Sechin, an early ceremonial center in the Casma Valley, depict armed conflict, S. and T. Pozorski (1987:79–82) have proposed that the murals were added to the monument in the late Early Horizon, not during the early phases of construction in the Initial period. Settlement pattern data from the Moche Valley indicate little or no warfare or raiding occurred in the Initial period and Early Horizon. In contrast to later periods, few sites are located in defensive settings and no fortifications were present (Billman 1996:181–183). Initial period leaders in the Moche Valley and probably elsewhere on the coast apparently did not control standing armies or mobilize groups for military actions. Ultimately, armed conflicted with less-centralized highland groups may have caused the demise of these early centralized polities on the coast at the close of the Early Horizon, ca. 400 BC (Billman 1996, 1997, 1999; S. Pozorski and T. Pozorski 1987). The onset of endemic warfare may have started a new cycle of political formation that resulted in the development of centralized, conquest-oriented states, such as the Moche state (Billman 1996, 1997, 1999).

EXPLAINING THE TIMING AND TEMPO OF PRISTINE CENTRALIZED POLITIES ON THE CENTRAL ANDEAN COAST

Why did pristine centralized political organizations develop along a 500-km section of the central Andean coastline around 1800 BC, while other areas of the central Andean coastline and the adjacent highlands did not develop such institutions until much later? Given the relatively similar environmental conditions along this 1800-km stretch of coast—a productive marine environment and irrigable land—why did pristine political formation not occur uniformly across the length of the coast, or in evenly spaced intervals, or randomly? Why one specific area of early pristine development? Further, why did political centralization take so long to spread out of the core area to the remainder of the coastline?

One possible answer to these questions is that pristine political centralization was a highly complex process involving the interaction of numerous variables. Consequently, any specific historical trajectory is unpredictable. Chaos theory has demonstrated that even relatively simple interactions between a few variables can produce unpredictable results, particularly if one or more of the variable relationships are nonlinear (Gleick 1988). For instance, simulations of weather involving the interactions of solar heat, gases, gravity, and friction (the lone nonlinear relationship in the simulation) have revealed that very slight changes in initial conditions produce highly chaotic, unpredictable results (Gleick 1988). Even if the formation of pristine centralized polities involved the interaction of a few behavioral and environmental variables, unpredictable historical trajectories seem a strong possibility, particularly given that several of the possible behavioral variables involved in the process, such as risk minimization, cost minimization, population growth, and status striving, are likely nonlinear.

Nonlinear relationships involve discontinuous change (sudden shifts between two conditions) or exponential rates of change. An extended discussion of the relationship between levels of societal violence and risk perception illustrates how nonlinear relationships can cause sudden, seemingly unpredictable shifts in the trajectory of sociopolitical evolution. Presumably people usually seek to minimize their risk of injury and death from violent activity. This type of risk minimization behavior probably plays an important role in political change, since security and self-defense are important concerns of both leaders and followers. In essence, leaders and followers seek to preserve and protect their families and themselves.

One's perception of risk is central to this type of risk minimization. In order to minimize risk one must first perceive risks, assess their level of danger, and then formulate a strategy of risk minimization. Undoubtedly, one of the main factors influencing one's perceptions of risk is the level of societal violence. If an individual perceives an increase in societal violence, then that individual is likely to change his or her risk minimization strategy. There is some evidence from contemporary and prehistoric societies that some societies have cycled in and out of periods of violence, often shifting suddenly and unpredictably between periods of relative peace and endemic warfare. These cycles of violence and peace may be related to the nonlinear relationship between levels of violence and risk perception.

With low levels of social violence, one's perception of the risks involved in participation in violent activities may dampen the level of social violence. If a person does not feel threatened by violent activity, then a person is probably less likely to engage in violence. The risks of injury or death would outweigh potential gains from engaging in violence. This, of course, is not true for everyone, but is likely true for most people in most societies when levels of violence are low.

As the level of violence increases, perception of risk changes. At some point, a critical mass may be reached and the perception of risk may swing dramatically in favor of participation in violent activity. For instance, in Lebanon, Somalia, Bosnia, or any number of other contemporary venues, the level of violence increased to the point that carrying on daily activities—taking the tram to work, relaxing in a sidewalk café, or shopping in the market—seemed to become no more risky than taking up arms and participating in the conflict. In other words, joining or supporting an army or militia may actually have seemed as safe or safer than going about one's business as an unarmed civilian. When the perception of risk changes in favor of participation in violence, violence can escalate rapidly, seeming to spin out of control.

Contemporary history suggests that eventually violence can escalate to the point that the perception of risk can shift suddenly back in favor of peace and reconciliation. Perhaps this is because at some point the perceived risks of violence become greater than the perceived risks of pursuing peace. There are several possible examples of this phenomenon in our contemporary world. After a generation of endemic violence and social chaos in Lebanon, the civil war suddenly and inexplicably came to an end. Ethnic and religious factions that were mortal enemies just a few years ago now interact peacefully on personal and political levels. Similarly, incidences of inner-city gang violence and murder in the United States, which had escalated rapidly without an end in sight in the 1980s and early 1990s, have dropped dramatically in the last few years. What this discussion illustrates is that risk minimization and risk perception may have a nonlinear relationship, and further that this nonlinear relation may be partially responsible for sudden and unpredictable shifts between war and peace.

Risk minimization is further complicated by other factors. For instance, the relationship between violence and risk perception is even more complex in social situations where classes, political factions, or other social groupings exist that have highly divergent perceptions of risk. For instance, leaders who start wars face a different risk equation than the people who have to fight in the wars. In addition, levels of societal violence, which dictate risk perceptions, are probably related to environmental stress, social stress, and status striving. This linkage of several nonlinear relationships in a chain of causality creates the potential for highly chaotic, unpredictable outcomes. Very slight changes at the start of the chain can be amplified by the nonlinear relationships, creating sudden unpredictable changes at the other end of the chain. Because periods of conflict create opportunities for leaders to develop military power bases, one of the keys to understanding conditions that promote political centralization are stress–violence, status striving–violence, and violence–risk perception relationships. What the above discussion suggests is that a slight change in environmental stress (or social stress or status striving) can cause a change in the level of violence. If violence reaches a threshold level, perception of risk can radically change in favor of participation or support of violent activity, resulting in a period of extreme violence. The shift to endemic conflict, in turn, can radically alter the power base of political leaders and leader–follower relationships, leading to sudden changes in the trajectory of political evolution. A slight initial perturbation can move through this series of nonlinear relationships causing sudden, unpredictable change at the end of the chain of causality. Given the complexity of just this one nonlinear behavioral relationship, even without the compounding complexity of divergent social groups and social statuses, is there any hope of understanding the process of pristine political centralization, except as a specific event resulting from a specific, unpredictable historical trajectory?

The notion that simple interactions can produce chaotic results is but one half of the lesson of chaos theory. Chaos theory also demonstrates that complex interactions can produce predictable results and that systems can cycle in and out of chaotic behavior. The very narrow temporal and spatial window of pristine state formation in human prehistory suggests that the formation of pristine centralized polities may be the predictable result of the interaction of a few simple relationships, although subsequent secondary state formation trajectories may display chaotic behavior.

The central Andean coastline serves as a case in point. Although predicting the specific valleys in which the process of pristine formation would occur may not be possible, the process at the regional level may be explainable in terms of a few relationships. At the regional level the process appears to be quite well patterned, with pristine polities developing only along a narrow section of the coast. A possible explanation for the pattern of development may be found in the frequency of El Niño events and the availability of arable land near the coast. El Niño events are less frequent the farther south one moves along the coast. El Niños occur every few years on the north coast, while on the south coast of Peru these events are rare. The effect of this is that marine resources are more stable and dependable the farther south a settlement is located. The quantity of irrigable land near the coast follows an opposite pattern. As one moves south along the coast, the coastal plain narrows and the quantity of irrigable land within easy walking distance from the ocean decreases. From the Ica Valley south, the foothills of the Andes begin at the ocean. As a result, irrigable land lies well inland from the coast except for a few small isolated pockets of bottomland.

The result of these two inversely related trends is that the optimal mix of relatively stable marine resources and relatively abundant coastal irrigable land is on the central section of the central Andean coast, between the Casma and Chillón Valleys. Although the north coast of Peru contains large quantities of coastal arable land, marine resources are far less stable than on the south coast. South coast valleys have dependable marine resources but lack significant quantities of arable land near the coast. The Chillón Valley and other neighboring valleys, midway between these two trends, have both abundant coastal arable land and relatively stable marine resources, apparently producing optimal conditions for the formation of early pristine centralized polities. In contrast, frequent marine perturbations or limited quantities of agricultural land may have inhibited the initial development and spread of political centralization on the south and north coast.

In essence, three features of the central Andean coastline may have strongly influenced the timing and tempo of the emergence of early pristine of polities in the region: (1) the opportunities for the sustainable surplus production created by abundant marine resources and highly productive irrigable land, (2) the opportunities for monopolistic control of production created by irrigation agriculture, and (3) the periodic downturns in production created by El Niños and droughts. As previously noted, the first two characteristics—sustainable surplus production and opportunities for economic control—are crucial for the development of economic power bases by aspiring leaders. The third characteristic—environmental perturbations—promotes the development of both economic and ideological power bases. As previously noted, periodic booms and busts create opportunities for well-positioned individuals or groups to develop debtor relationships with less well off or less lucky groups. By amassing debts, which could be repaid in labor or goods at a later date, aspiring leaders could develop a fund of surplus goods and labor that could be used to finance political activities and extend their political influence and control. In essence, debts could be developed as an economic power base. Unlike economic power, the conditions that promote the growth of ideological power are poorly understood and are somewhat of a weak link in theories of political evolution. Nonetheless, boom and bust cycles do appear to be one of the factors that promote ideological power by allowing leaders to rally groups to a cause and claim credit for the return to good times.

In contrast to economic and ideological power, the complex nonlinear relationships between stress, status striving, violence, risk perception, and the onset of endemic warfare suggest that the circumstances that promote the development of a military power base may be the least predictable of the three sources of political power. Cases of pristine political formation that relied on the development of military power may have been the most chaotic and least predictable. Surprisingly, military power and conflict do not appear to have played a significant role in this case study, although subsequent research may disprove this notion. Conflict and military power clearly were important in the formation of subsequent Andean states, most notably the Moche and Inka states.

If these three environmental conditions (abundance, control, and perturbations) were responsible for the evolution of political centralization on the central Andean coast, then what this case study illustrates is that there may be certain threshold values for these conditions that promote pristine political formation. Too great or too low a frequency of environmental perturbations and the development of centralized political systems was inhibited. Likewise, the potential for a certain minimum level of sustainable surplus production and economic control must be present. Optimal levels for these three variables may have been met on only a small portion of the central Andean coast, creating potential for development of some the first centralized polities in the New World out of the interaction of human behavior traits—such as risk minimization, cost minimization, and status striving—and key environmental variables.

ACKNOWLEDGMENTS

I would like to thank Jonathan Haas for organizaing the Leaders to Rulers Conference at the Field Museum and for his comments on an earlier draft of this chapter. Special thanks go to the conference participants for making the conference an exciting and stimulating event. Illustrations for this chapter were drafted by Chris Rodning.

Chapter **10**

"Who Was King? Who Was Not King?"
Social Group Composition and Competition in Early Mesopotamian State Societies

GIL STEIN

INTRODUCTION

The last two decades of archaeological and textual research have documented tremendous diversity in the ways that Greater Mesopotamian complex societies constituted themselves as polities (Fig. 1). This increasingly representative database, combined with the use of more processually oriented models of social action, have led to a gradual shift in research perspectives from a "top-down" emphasis on managerial structure toward a "bottom-up" perspective on the organization of Mesopotamian chiefdoms and states (see, e.g., Stein 1994a; Yoffee 1995). The traditional structural approach treated Mesopotamian complex societies as homogeneous, highly centralized entities whose urbanized governing institutions defined and controlled virtually every aspect of economic, political, and social life. This largely implicit view derived from the historic emphases of Near Eastern archaeology and philology. For over a century, archaeologists had concen-

---

GIL STEIN • Department of Anthropology, Northwestern University, Evanston, Illinois 60208.

From Leaders to Rulers, edited by Jonathan Haas. Kluwer Academic/Plenum Publishers, New York, 2001.

Figure 1. Map of Mesopotamia.

SOCIAL GROUP COMPOSITION AND COMPETITION 207

trated on the excavation of monumental public buildings such as palaces and temples in major urban sites (see, e.g., Lloyd 1980). Similarly, Assyriologists tended to view the cuneiform archives of these centralized institutions as complete and representative records of the full range of activities, institutions, and interest groups in Mesopotamian society. This urban, elite-oriented focus was perfectly understandable, given the fact that Mesopotamia is the earliest known and best-documented ancient urban society.

However, a recent series of complementary developments in Assyriology, archaeology, and anthropological models of complexity all have contributed to the emergence of a fundamentally different analytical framework. Instead of viewing Mesopotamian complex societies as tightly integrated, highly centralized polities, researchers have come to focus on the roles of heterogeneity, contingency, and competition among different sectors and interest groups as critical factors whose interaction defined the fabric of Mesopotamian society. The organizational forms of Mesopotamian complex societies emerged through the dynamic interaction of partly competing, partly cooperating groups or institutional spheres at different levels of social inclusiveness (see also Mann 1986:1–33). Reconstructions of these polities must combine both archaeological and textual materials as complementary sources of information.

One of the most positive developments connected with the increasingly historical orientation in recent research on ancient literate states has been the explicit integration of textual and archaeological data, while fully recognizing that each of these complementary sources of information has its own sampling and interpretive problems. To a large extent this has become possible because archaeologists, philologists, and ancient historians have greatly improved our understanding of the social context and political uses of literacy in early state societies such as Egypt and Mesopotamia (see, e.g., Baines and Yoffee 1998).

Mesopotamian state societies play a central (although often implicit) role in anthropological theories of complex societies as a result of their being the earliest known polities of this type in the world, and more importantly, due to their extensive written record of cuneiform texts that preserve otherwise inaccessible details about political history, ideology, and economic organization. Mesopotamia was the first society to invent writing. The early polities of this region used a cuneiform script to write two main languages—Sumerian (a linguistic isolate) and Akkadian (a Semitic language ancestral to Aramaic, Hebrew, and Arabic). The Mesopotamians did most of their writing on clay tablets, which means that literally hundreds of thousands of documents have survived and are available for analysis.

However, writing has been a mixed blessing for anthropological archaeologists studying early Mesopotamian state societies. On the one hand, written documents provide information on aspects of early state organization and ideology that are not preserved and cannot be reconstructed from nontextual or archaeological evidence alone. At the same time, textual evidence can be profoundly misleading, especially when we are attempting to reconstruct the political economy of early state societies. The first cities and state societies developed early in the fourth millennium BC, during the Uruk period. Shortly afterward, writing was invented in the Late Uruk period, probably some time between 3400 and 3200 BC (Green 1981; Nissen 1985, 1986; Nissen et al. 1993) (Fig. 2). The earliest uses of writing were administrative; Uruk texts are almost all records of the administration of economic activities. Only in later periods did the use of writing expand to more overtly

MESOPOTAMIAN CHRONOLOGY

Period	Dates	Comments
Hellenistic	330-143 BC	
Median/Persian	539-330 BC	
Neo-Babylonian	625-539 BC	
Late Assyrian	911-612 BC	
Middle Assyrian Middle Babylonian	1363-1076 BC	
Kassite	1415-1154 BC	
Old Babylonian	1812-1595 BC	
Isin Larsa (South)	2025-1736 BC	
Old Assyrian (North)	1900-1740 BC	Trading colonies in Anatolia
(collapse) ca.	2000-1800 BC	"Amorite Invasions"
Ur III Dynasty	2112-2004 BC	
(collapse)	2193-2100 BC -	"Gutian Invasions","period of Gudea"
Akkadian Empire	2334-2193 BC	Sargon, Naram-Sin- First empire
Early Dynastic I-III	2900-2334 BC	Sumerian city-states, Ur royal cemetery
Jemdet Nasr	3100-2900 BC	Spread of urbanism & kingship
Uruk	4000-3100BC	First cities, states, writing, & colonies
Ubaid III-IV	4500-4000 BC	Spread of Ubaid to North Mesopotamia
Ubaid I-II	5500-4500	Complex societies integrated through ritual
Samarra	5500-5000 BC	Earliest canal irrigation in Mesopotamia
Halaf	5300-4500 BC	

Figure 2. Chronology of Mesopotamian state societies.

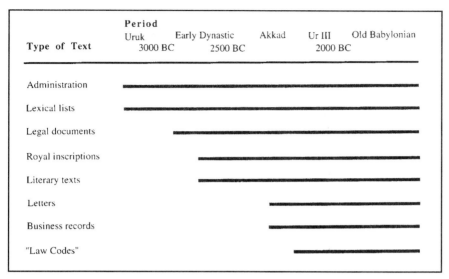

Figure 3. The expansion of the uses of writing in Mesopotamia.

ideological uses such as letters, literary/historical texts, ritual texts, and royal inscriptions (Postgate 1992:66) (Fig. 3).

As a result, there are two main sources of bias in the use of textual data to reconstruct the political economy of the Uruk state and its successors in the third millennium BC. The first of these is the sampling bias that arises from using administrative texts from palace or temple contexts to reconstruct power relations in the society as a whole. The second bias arises after the Uruk period, once writing started to be used to record literary, historical, ritual, and royal discourses. At this point, we are faced with the bias of propaganda. Although state and elite claims to authority are valuable in their own right as a way to reconstruct ideologies of rule, we still need to determine the extent to which these ideological assertions accurately reflect the complex and shifting power relationships of the different groups that made up Mesopotamian society: "In all instances, we have to keep foremost in our minds that even strictly historiographic documents are literary texts and that they manipulate the evidence, consciously or not, for specific political and artistic purposes" (Oppenheim 1977:144). Thus, somewhat paradoxically, as written records become increasingly available as sources of information about Mesopotamian state societies, so too do the interpretive difficulties, sampling problems, and propagandistic biases increasingly affect our models of political economy.

These administrative and ideological biases—the first accidentally and the second often deliberately—present us with a skewed picture that overemphasizes the economic, military, political, and ideological power of the centralized ruling apparatus, or the "great institutions" of ancient Mesopotamian states. The textual data can tell us much about the ideology of centralization, but they provide misleading information about the actual operation of ancient Mesopotamian states.

In assessing the extent and effects of these biases, we must remember that literacy was limited to less than 1% of the Mesopotamian population (Larsen 1989); as a result, textual data tell us little or nothing about most other social sectors, such as urban commoners, independent craft specialists (Stein and Blackman 1993), the rural sector, and pastoral nomads. Women as well are generally excluded or marginalized in written records (with the occasional exception of female dependents on the ration lists of palaces and temples or a small number of high-status women connected with either ruling families or temple priesthoods) (see, e.g., Durand 1987; Pollock 1991b; Winter 1987; Wright, 1996). In almost all cases, information about these textually underrepresented groups can come only from the archaeological record. Integration of archaeological and textual data also can highlight disjunctions between the state's ideologically based claims to power (reconstructed from texts) and archaeologically detectable patterns of economic organization in the society as a whole.

The title of this chapter is a quote from a document called the "Sumerian King List," one of the most important literary texts in Mesopotamian history. It lists all the kings of Mesopotamia for a period of more than 1000 years, from the time "when kingship descended from heaven" in the beginning of the Early Dynastic period (ca. 3000–2900 BC) down into the second millennium BC. This frequently recopied text is not just a list, but also is a charter of legitimacy, or more accurately, of claims to legitimacy (Michalowski 1984). The Mesopotamians believed that legitimate kingship only resided in one place at any one time. However, at the end of the third millennium BC, the structure of Mesopotamian society had broken down so badly into competing warring polities and claimants to power that instead of the normal royal formula, the scribes were forced to throw up their hands figuratively and ask, "Who was king? Who was not king?" The scribes were, of course, confronted with the contradiction between the normative, propagandistic assertions of a universal royal ideology on the one hand and the harsh reality of conflicting claims to authority in a fragmented social landscape. This brief quotation highlights the fact that much of the rich textual record for ancient Mesopotamian states makes claims to legitimacy, power, and authority, but cannot fully deliver on that promise (see also Yoffee 1995:281–285).

SOCIAL GROUP COMPOSITION AND COMPETITION

CHANGING PERSPECTIVES ON THE ORGANIZATION OF MESOPOTAMIAN STATES

The first state societies developed in Mesopotamia during the Uruk period, sometime between 3800 and 3600 BC. By about 2700 BC, in the Early Dynastic period, we see the real flowering of Sumerian culture and the development of Sumerian writing to a point where the texts become readable, important historical sources for the reconstruction of political economy in this period. Five major social trends characterized the state societies of Early Dynastic Mesopotamia.

Widespread Urbanism

In the previous Uruk period, the city of Warka was the major urban center in Mesopotamia, standing at the top of a four level settlement hierarchy (Adams 1981:71; Johnson 1973; Wright and Johnson 1975). In the Jemdet Nasr and Early Dynastic periods, urbanization spread widely so that there were numerous large city states all across southern Mesopotamia, notably at Umma, Lagash (al-Hiba), Uruk, Ur, Shuruppak, Kish, and Nippur (Adams 1981).

Endemic Warfare

The Sumerian city states of the Early Dynastic period were constantly at war with each other, as we can see from massive defenses, numerous finds of weapons, and depictions of warfare (Charvat, 1982; Gelb 1973; Nissen 1972; Winter 1985).

Emergence of Kingship

Although icononographic evidence suggests that kingship first developed in the Uruk period, the growth and spread of powerful royal dynasties in the Early Dynastic period is probably connected with their role as military leaders in the warfare between rival city states (Charvat 1982; Michalowski 1984; Moorey 1978; Oppenheim, 1977:98; Schmandt-Besserat 1993).

Pronounced Social Stratification

Kings were at the top of a very complex social hierarchy. Some of our best evidence for this comes from the royal cemetery of Ur where we have hundreds of graves of commoners and 16 lavish tombs, filled with gold, silver, lapis lazuli, and large-scale human sacrifices of the retainers who accompanied the kings and queens of Ur into the next world (Moorey 1977; Pollock 1991a; Woolley 1934).

Increasing Use of Writing

Although writing was invented at the end of the Uruk period, the extremely abbreviated nature of these earliest pictographic texts makes it difficult to use them for anything more than reconstructions of recording and bookkeeping systems (Nissen et al. 1993). Writing did not become widespread until the Early Dynastic period. Of course, "widespread" is a relative term. Writing was a very specialized technology that was used in every Sumerian city but only by a tiny group of scribes in a very limited number of social contexts. We have some royal inscriptions, but most of the documents from this period are economic records from large temple archives. They show temples as vast, wealthy, and powerful organizations, owning massive amounts of agricultural land, controlling the labor of hundreds or even thousands of dependent laborers (Oppenheim 1977; Postgate 1992).

These documents provide a wealth of data about the complex political economy of Early Dynastic Mesopotamia, but this information has come at a price. Once written documents are present, the texts tend to drive the research agenda; they define our perceptions of Mesopotamian society, and structure the entire way we organize our research. The most seductive aspect of the textual record is that it gives us the illusion that we know more than we actually do about Mesopotamian society. When literally thousands of documents are available, one can easily fall into the trap of assuming that the written record is an accurate and complete record of Mesopotamian society. Textually based biases have had an enormous and not always useful effect on the models that people have suggested to describe and explain the political economy of complex societies in Mesopotamia.

EARLIER MODELS OF CENTRALIZED MESOPOTAMIAN STATES

One of the earliest and most influential models of early Mesopotamian state organization was the textually based concept of the earliest Mesopotamian polity as a highly centralized, monolithic "temple–state." This reconstruction was proposed in 1931 by Father Anton Deimel (Deimel 1931). The temple state model was based on the approximately 1600 texts from the archive of the Bau temple at Tello (ancient Girsu) in the Sumerian city state of Lagash. These documents described temple control over land, agricultural laborers, craft workers, and the collection of tithes. Deimel calculated the extent of the Bau temples land holdings and estimated that the temple owned all the land in the territory of Lagash. On that basis he suggested that Early Dynastic Sumer was a series of highly centralized theocracies where temple priests controlled the entire state. The temple–state model was widely embraced by both Assyriologists and archaeologists for over 40 years (Falkenstein 1974). Although this reconstruction of Early Dynastic Mesopotamian states was finally critiqued and discredited by the work of Assyriologists such as Ignace Gelb and I. M. Diakonoff in the late 1960s and early 1970s (Diakonoff 1974; Gelb 1969; Foster 1981), it still survives in much of the popular literature and even in nonspecialist archaeological syntheses of early Mesopotamian history.

During the 1970s and 1980s, a major debate emerged concerning the nature of the state and its implications for the study of early civilizations such as Mesopotamia: Should one focus on the states as integrated, specialized, political-administrative entities or should one study states as whole but heterogeneous societies? The two approaches worked from different basic principles and led to fundamental differences in analytical focus.

The integrative model of states as homeostatic systems drew on ecology and systems theory to view a given state society such as Mesopotamia as a single integrated system made up of different specialized subsystems (Flannery 1972). The state or government was seen as regulating the flow of goods, services, and information among the lower order subsystems (see, e.g., Johnson 1973, 1978, 1982; Redman 1978; Wright 1969, 1977, 1978; Wright and Johnson 1975). This view assumed that the administrative and decision-making subsystems controlled all major economic and political aspects of an integrated social system. This centralized, control-oriented systems model fit well with the picture from the textual data of vast temple and palace bureaucracies regulating large amounts of labor, land, and capital. The focus on regulatory mechanisms, decision making, and administration was so strong that it was used to define the state as the essential component of society due to its managerial and coordinating function. The view of states as administrative systems was explicitly aimed at developing cross-culturally valid general principles to explain the development of social complexity. In this model, the key research problem was to explain what perturbations to the system required it to adapt and maintain homeostasis by developing increasingly complex and hierarchical decision-making organization to continue regulating all the subsystems of society effectively.

In contrast to the integrative systems model of highly centralized Mesopotamian states, more holistic models had a much stronger historical or particularistic element, concerned with understanding the specific factors that were critical in the development of Mesopotamia as a complex society (Adams 1981:76–77). This approach focused on trying to document the broad social transformations that affected all of Mesopotamian society, specifically the processes of urbanization and the development of social stratification, as a substitute for the exclusive focus on administration and bureaucracy in the central state institutions. This perspective saw urbanism as not just a demographic phenomenon but also a fundamental change in social organization into communities that were larger, more specialized, hierarchical, and diverse than any other social grouping that had ever existed on the planet. This historically oriented approach focused on economic and social evidence to trace the development of urbanization and social stratification.

STATES AS HETEROGENEOUS SOCIAL NETWORKS

Building on this shift, in the last decade a heterogeneous, conflict-based model of state societies has developed (Brumfiel 1989, 1992, 1994; Crumley 1995a; Ehrenreich et al. 1995; Gailey and Patterson 1987; Gledhill et al. 1988; Mann 1986; Stein 1998); in the case of Mesopotamia, this can be seen as a shift from attempts to explain state origins toward a concern with "organizational dynamics," an attempt to figure out how these societies actually functioned (Stein and Rothman 1994). This model rejects the idea that societies are tightly integrated, regulated adaptive systems. Instead, it is probably more realistic to see societies as "fuzzy networks." I use the word "network" deliberately in order to emphasize that there is a larger whole composed of interconnected and interacting parts, but these parts need not be integrated in the kind of centralized rigid structure implied by the term "system." Networks are looser and not necessarily coercive. Social networks are also "fuzzy" in the sense that they do not have clear, uniform boundaries. Instead, we can identify these networks based on frequency of interaction, so that it is very clear what lies at the heart of the network, but less so at the edges.

This movement away from seeing state societies as self-regulating systems carries with it a shift in the way people use the concept of decision making. The earlier evolutionary systems model dealt with decision making as a monopoly of the administrative subsystem of society. More recent views emphasize the roles of human intentionality, agency, and strategizing in all sectors of society, not just chiefly or state elites. This means that societies are not composed of subsystems; they are composed of thinking people who are rational in terms of their own cultural values and social position. These people have goals and organize themselves into different groups to carry out those often conflicting aims.

SOCIAL GROUP COMPOSITION AND COMPETITION   215

Four main principles are especially important in this heterogeneous model of complex societies:

1. Societies are not unitary systems; instead, they are heterogeneous composites, made up of numerous overlapping and interacting groups.
2. Different groups in the same society can have completely different and often opposing goals. The scope of these groups extends beyond the hierarchically organized classes in the Marxian sense to include a broader range of social, cultural, economic, and kin-based units.
3. The goals and strategies of different groups are situationally contingent and can change over time so that the society will have a shifting mix of cooperation and competition among different groups.
4. Decisions made in terms of the short-term goals of these groups can have completely unintended long-term consequences. This is important because it implies that even though people have intentionality and agency, this does *not* mean that everything we see in a society was deliberately engineered to turn out that way. Major change is often unintentional, even in state societies that are nominally under highly centralized "absolute" rule.

Archaeological models recognize the limits of central power in chiefdoms (see, e.g., Earle 1991a; Spencer 1987; Wright 1984) but generally have an implicit view of the state as being all powerful. By counterposing chiefdoms and states as discrete categories, archaeologists have tended perhaps to overestimate the degree of centralization and power of the latter. Analyses of complexity and integration in Old and New World complex societies suggest that there is marked cross-cultural variation in both the administrative structure and the ideological, economic, or political strategies used by elite groups as instruments of rule (Blanton et al. 1996; Earle 1991a; Feinman 1995). However, despite this diversity, virtually all complex societies share an inherent tension between the opportunistic, centripetal tendencies of a chiefdom or state as a political institution/governmental structure (Adams 1981:76–77; Wright 1977) and the centrifugal tendencies of the other sectors of the broader society (see, e.g., Adams 1978, 1982; Yoffee 1979:21). Consequently, instead of viewing states as all-powerful, homogeneous entities, it is probably more accurate to characterize them as organizations operating within a social environment that for a variety of reasons they only partially control.

This alternative model sees Mesopotamian states as generally operating suboptimally, due to their inability to attain or maintain their desired levels of power, authority, legitimacy, and control over the wider society around them. When Mesopotamian states pursued maximizing strategies aimed at extracting large, consistent surpluses from the countryside (Adams 1978), these attempts tended to be short-lived, unstable, and vulnerable to collapse from both internal and external stresses (Wilkinson 1994; Wilkinson and Tucker 1995). In particular, they often resulted in environmental damage that limited the very agricultural potential they wished to enhance (Gibson 1974; Jacobsen 1982; but see also Powell 1985). As a result, the centralized authorities of these polities would more commonly have been forced to eschew optimizing strategies in favor of what Simon (1959) called "satisficing" strategies. The state's satisficing strategies would have focused on survival, self-perpetuation, or social reproduction, while attempting to expand or increase its control on an opportunistic basis, whenever local political circumstances permit it.

State institutions faced limits on the exercise of their powers both within the urban centers and in their attempts to extend control over the countryside. Some sectors of society, such as nomads, lay completely outside the purview of the state's "core concerns." Other sectors of society may have fallen partially within the boundaries of the state's core concerns, but this does not necessarily mean that governmental institutions were able to control them. As just one sector within an unevenly integrated social landscape, the Mesopotamian state institutions were often forced to duplicate the subsistence and craft activities of other sectors, in order to provide for their own core concerns or survival needs. For example, the palace sector appears to have operated as a large, autonomous household, producing its own crops, animals, and utilitarian craft goods through its control over "attached" specialists (Stein and Blackman 1993; Stein 1996), rather than obtaining them through exchange with independent craft specialists or food-producing village communities. This duplication of economic activities and institutions reflects poor integration of the different sectors of society. It is essentially a kind of "dual economy" characterized by a high degree of sectorial autonomy. One pole of this dual economy, that of the centralized urban institutions, is prominent in both the textual record and archaeological evidence from large-scale public buildings in urban contexts. By contrast, the villagers, nomads, independent craft specialists, and other urban commoners who form the opposite pole of this dual economy are almost invisible in the cuneiform documents. These latter groups generally emerge only through archaeological research focused on village communities, urban residential quarters, craft production, exchange systems, and survey work to elucidate regional economic patterns.

What this means for archaeologists is that excavated sites and mapped settlement patterns are not the remnants of an integrated system with a unitary logic and a preconceived pattern. Instead, what we are finding is a heterogeneous mixture: the observed outcome that we call "Mesopotamian

SOCIAL GROUP COMPOSITION AND COMPETITION

society" is in fact the combined result of the interaction and jockeying for power among different often competing groups, each with its own goals and decision-making strategies. We cannot begin to get an accurate view of how these polities functioned unless we take into account the large proportion of society that appears only fleetingly or not at all in the cuneiform texts.

"SECTORS" IN EARLY MESOPOTAMIAN STATE SOCIETIES

Based on a combination of textual sources and archaeological data, we can identify several main interacting "sectors" or groups in Mesopotamian society. The most visible parts of Mesopotamian society are the "great institutions" or "great organizations" (Oppenheim 1977:95–109), a general term referring to two overlapping but often competing foci of power: the royal palace and the temple sector. These are the parts of society that left behind the monumental public buildings that predominate in excavations of Mesopotamian cities. The thousands of cuneiform tablets kept by the bureaucrats in these institutions give us detailed information about their economic organization. However, the great institutions are simply the most visible parts of a complex society that consisted of many overlapping and often conflicting social sectors.

The Palace

The palace sector consisted of the king, his royal household, and their extensive economic holdings of irrigated fields, orchards, flocks of sheep, and cattle, along with extensive stores of food, craft goods, and precious raw materials (see, e.g., Archi 1985; Dolce 1988; Matthiae 1978; Pinnock 1988). The royal household consisted of the king's extended family, administrative officials, craft specialists, a standing army, serfs or "semifree" workers (see below), and slaves. Sargon of Akkad, the founder of the Akkadian empire, boasted that he fed more than 5400 palace dependents every day. In addition to enormous economic power, the authority and power of the palace sector rested on the king's role as a war leader and on his control over a large, well equipped standing army (Garelli 1974; Oppenheim, 1977:98–105).

The Temple

Every Mesopotamian city state had several large temples, each dedicated to a different god and run by a professional class of priests. These organizations were considered to be "households" belonging to each god, so they were as much economic as they were ritual institutions. The temples and their priesthoods controlled many of the same resources as the palace—land, flocks, slaves, semifree gurush workers, and craft workshops. The temples also collected tithes from their worshippers, and acted as banks, providing loans of money or grain (Lipinski 1979; Postgate 1992:109–136.

Large Estates

Less well documented than the palace or the temple were the great families of wealthy Mesopotamians who were not directly part of the palace or temple sectors but also controlled large private holdings of fields, date palm orchards, flocks and other forms of wealth (Adams 1966:106–107).

Craft Specialists

Craft specialists were organized in several different ways in Mesopotamian society, so it is not really accurate to treat them as a uniform group. One of the biggest distinctions would be between "attached" and "independent" specialists. Attached craft producers were employed by or dependent on either the royal household or a temple (Maekawa 1980; Tjumenev 1969; Zaccagnini 1983). The great institutions controlled the products made by these specialists and in return would support them with rations of food, beer, and clothing. They also would provide the attached craft specialists with the tools and raw materials needed for their profession. We know of attached specialists because they appear in the textual record on the lists of people receiving rations and supplies from the palace or the temple storerooms. There also were independent craft specialists, who worked outside of the purview of the great institutions (see below).

Urban Commoners

Although they are invisible in the textual record (until the second millennium BC; see, e.g., Stone 1987), surface survey and limited excavations suggest that at least half of the Sumerian population lived in cities during the third millennium BC (Adams 1981:138; Postgate 1994). These people were probably for the most part agriculturalists, farming the lands immediately around the cities (Adams 1972), although a small merchant class existed as well, at least as early as the Akkadian period (Foster 1977; Westenhotz 1984). Archaeological evidence suggests that urban commoner populations included independent craft specialists as well. We know that the citizens of

Mesopotamian cities had political representation and power through the institution of the assembly or council of elders. In addition to their right to conduct real estate transactions and make legal decisions, these assemblies were powerful enough to successfully claim legal and fiscal privileges from the palace (Oppenheim 1977:112; Jacobsen 1943, 1957; Yoffee 1995).

"Semifree" Individuals

One very large social group in Mesopotamia were those called Gurush/ Geme. These were people reduced by debt or poverty to the point that they became indentured to the great institutions, for whom they worked as more or less unskilled agricultural laborers with restricted freedom, in return for rations of food and clothing (Davidovic 1987; Diakonoff 1976; Gelb 1965; Maekawa 1987; Milano 1987).

Slaves

Slaves were also common in Mesopotamian society. Probably most slaves worked in domestic service (Diakonoff 1976).

Villagers

The village sector of Mesopotamian society almost never appears in the textual record. Yet we estimate the number and size of settlements found by archaeological surveys that up to 50% of the Sumerian sedentary population lived in villages. These groups were apparently organized as autonomous, kin-based landholding communities (Diakonoff 1975). These communities appear fleetingly in the textual record through finds of *kudurrus* (inscribed boundary stones), which record royal purchases of land from villages that seem to have held corporate ownership over this primary agricultural resource (Gelb 1969; Gelb et al.1991).

Nomads

Pastoral nomadic groups are the great invisible sector of Mesopotamian society (Cribb 1990). Nomads show up only in the most indirect way in the textual record (Adams 1974). We know they were there from periodic references to Mesopotamian kings either fighting them or making treaties with them to allow for safe passage of the caravan trade (Kupper 1957). In periods of weakened centralized authority, nomadic groups would increase their raiding activities. Nomadic raiding could and did periodically bring about the downfall of Mesopotamian states (Postgate 1992:43). In addition to being an important military force to be reckoned with, the nomads were presumably also an important source of meat, dairy products, and wool to the village sector of Mesopotamia. Wool and textiles were some of the most important commodities in the Mesopotamian economy, so we would expect nomads to have been an economically powerful group. The line between nomads and sedentarists is not hard and fast. People are always moving from one status to the other. Based on analogy with modern groups, we can assume that rich nomads often became sedentary, while the poorest of the poor nomads may have been forced to become semifree gurush in order to survive. Similarly, villagers in Mesopotamia always kept animals in addition to their fields, as a kind of insurance. Under conditions of drought or if taxation by the cities became to great, these groups could become increasingly mobile, both to survive economically and to evade state control. In other words, the social boundaries in Mesopotamia were extremely fluid, so that even though cities, villages, and nomads were often in conflict, there also were periods of collaboration and much movement from one sector of society into the other.

These nine main social groups appear differentially in the textual record. We know the most about the palace and the temple sectors because those were the institutions that used scribes and writing. We know a bit about dependent social groups like attached craft specialists, semifree laborers, and slaves, but only through their appearance on the ration lists—the payrolls—of the great institutions. But the majority of the Mesopotamian population is for all intents and purposes textually invisible. The textual record of the temples and palaces gives the impression of a highly centralized bureaucratic system. But once we factor these textually underrepresented sectors back into the social equation, they fit very well into a heterogeneous model of Mesopotamian states in which the great institutions were important but not the only powerful or influential social actors. Textually invisible or underrepresented sectors limited the exercise of centralized state power both within the urban centers themselves and especially in their hinterlands.

The tension between centrifugal conflict and the centripetal push toward integration created in Greater Mesopotamia a dynamic system of parallel, partially overlapping, and often competing social spheres, whose modes of interaction varied widely in time and space (Stein 1994a). The temple and state "sectors" were the main centripetal forces of integration and centralization. These great institutions attempted to extend administrative, political, and economic control over the other social sectors in order to gain access to the surplus labor, craft goods, agricultural goods, and pastoral products that were critical for the survival and social reproduction of Mesopotamian elites. However, the centralizing strategies of elite institu-

SOCIAL GROUP COMPOSITION AND COMPETITION 221

tions were counterbalanced by opposing centrifugal strategies of autonomy and resistance in other "sectors" of Mesopotamian society. Inside the cities, the temples and palace sector often were in conflict. The palace and wealthy commoners pursued markedly different economic strategies (Rothman 1994). As noted above, the urban assemblies or councils of elders could and often did oppose palace decisions. However, the main centrifugal force in Mesopotamian society was the persistent opposition by the countryside to centralized urban control. The rural village and nomadic sectors " . . . show definite and often effective resistance not only against living in settlements of greater complexity than the village but also against the power—be it political, military, or fiscal—that an urban center was bound to exercise over them" (Oppenheim 1977:110).

THE ARCHAEOLOGY OF "INVISIBLE" SECTORS IN MESOPOTAMIAN STATE SOCIETIES

Archaeological fieldwork in Greater Mesopotamia has played a major role as a complementary line of evidence to clarify the social role of these textually underrepresented sectors or social spheres in Mesopotamian society. Broadscale surveys of the south Mesopotamian alluvium (Adams and Nissen 1972; Wright 1981), southeastern Iran (Johnson 1973, 1980, 1987; Wright and Johnson 1975), the north Syrian/North Mesopotamian steppe (Stein and Wattenmaker 1990; Wilkinson and Tucker 1995), and southeast Anatolia (Algaze 1989; Algaze et al. 1991; Whallon 1979; Wilkinson 1990) have documented the ways in which patterns of rural organization evolved in tandem with cycles of integration and collapse in Mesopotamian complex societies. Assyriologists, too, have begun to recognize the importance of survey to obtain a more representative picture of ancient Mesopotamian society (Brinkman 1984; Postgate 1986).

Excavations and surveys by Hole (1981) have focused on the role of ancient pastoral nomads in ancient Mesopotamian society. Excavations at sedentary village sites have produced some of the first hard evidence for rural productive systems in Mesopotamia (Wright 1969), Syria (Curvers and Schwartz 1990; Schwartz and Curvers 1992), and southeast Turkey (Algaze 1990; Stein 1987; Wattenmaker 1987). Concurrently, a series of anthropologically oriented excavations and intrasite analyses of urban centers have done much to document the spatial patterning of residential, economic, and administrative functions (Pollock et al. 1996; Stone and Zimansky 1994, 1995); neighborhood organization (Stone 1987); craft production (Blackman et al. 1993; Nicholas 1990); and the organization of urban food supplies (Zeder 1991).

Two case studies can show how archaeological analyses of textually invisible parts of society can both complement the written record, while showing that we need to recognize the heterogeneous nature of Mesopotamian society in order to understand how it worked. The following two case studies examine the regional organization of crafts and subsistence in urbanized state societies of Northern Mesopotamia in the third millennium BC. These examples show that the rural hinterlands of urbanized states pursued economic strategies that paralleled, differed from, and probably competed with the goals and strategies of the state institutions in the urban capitals of these polities.

CRAFT PRODUCTION AND "DUAL ECONOMIES" IN MESOPOTAMIA

In assessing the textual evidence for craft production, it is important to emphasize that the existing corpus of late fourth millennium texts is a nonrandom, nonrepresentative sample that records the transactions of a very limited subset of Mesopotamian society—the "great institutions"—the palaces and temples. Even when textual evidence is available from palace or temple archives, we must remember that these documents represent specialized, abbreviated records of highly specific types of transactions that took place within circumscribed administrative spheres. Thus, the fact that the texts might mention a particular craft or activity does not necessarily mean that the latter was a state monopoly. It is important to distinguish between the crafts and activities that the "great institutions" practiced along with everyone else in Mesopotamian society and those crafts or activities that the state actually controlled. It is more accurate to view the Mesopotamian state as consisting of several different social domains such as an "institutional sphere" consisting of the temples, the palaces, and what we can call "noninstitutional" urban and rural spheres, each pursuing a more or less similar range of core activities in tandem, with only minimal interaction or exchange of value among the three domains. While the state institutions asserted general claims to control over the territory of their urban-centered city states, in practice this control was variable, contingent, and often quite restricted.

The Mesopotamian temple or palace was in many ways analogous to an extremely large and wealthy household; as such, it retained, as maintenance staff, low-status dependent workers who provided the goods and services needed for the internal consumption and day-to-day operation of that institution (see Lipinski 1979; Powell 1987). The inclusion of these dependent workers in the archives of the palaces and temples in no way implies that the state monopolized control over these professions. Instead, these dependent workers merely duplicated, within the administrative sphere of the temple or palace, activities that were also taking place outside the purview of the great institutions in Mesopotamian society as a whole. These "duplicated" professions (and their products) apparently included a wide range of specialists such as farmers, herders, fishermen, woodcutters, carpenters, weavers, bakers, and possibly even potters as well, this last group producing utilitarian pottery for the internal consumption of the temple or palace.

From the Late Uruk period onward, Mesopotamia essentially had a "dual" craft economy (Fig. 4) consisting of both independent craft specialists operating in the community at large and attached craft specialists who were clients of the centralized institutions. The attached craft specialists themselves can be divided into two groups: those who produced the same goods as the independent specialists and those who manufactured high prestige goods.

The limited available evidence suggests that only those specialists who manufactured high prestige goods were actually under the complete or monopolistic control of the centralized state institutions in the fourth and third millennia BC. In Late Uruk Mesopotamia, gold, silver, copper (Moorey 1982:21–23), carnelian, polished stone vessels, and elaborate textiles appear to have been the main prestige goods whose display legitimized the

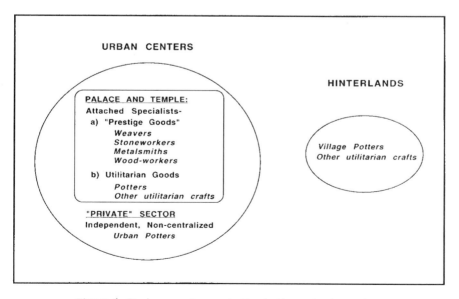

Figure 4. Dual economic organization in Mesopotamian society.

high status of the elite individuals or institutions who possessed them. By contrast, other goods such as utilitarian ceramics or low status textiles would have been produced both by attached palace specialists and by independent specialists operating outside the purview of the centralized institutions (Stein and Blackman 1993; Stein 1996).

Although they rarely appear in the textual records of the great institutions, we can infer the existence of independent specialists from archaeological discoveries of metalworking, weaving, and ceramic workshops in regular city neighborhoods or in the villages surrounding the cities. These people lacked institutional patronage and presumably exchanged their products with urban commoners, villagers, or nomads. Archaeological evidence from excavations in the major urban centers and surveys of their hinterland confirms the existence of a dual economy characterized by parallel urban and rural independent craft production.

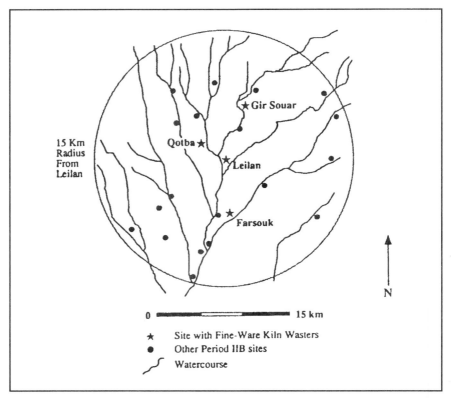

Figure 5. The urban center of Leilan (North Syria) and its hinterlands, showing locations of fine ware bowl mass production, based on finds of stacked kiln wasters.

At the city of Leilan in northwest Mesopotamia, the emergence of urbanism was accompanied by a major change in ceramic production, with the widespread distribution of open, simple rim, flat base, undecorated fine wares as the most common and characteristic ceramic form in the urbanized Period IIb (Schwartz 1988:40–42). High concentrations of fine ware wasters in the lower town of Leilan attest to the mass production of these ceramics within the urban center (Blackman et al. 1993; Senior and Weiss 1992; Stein and Blackman 1993).

Archaeological surveys of the Leilan region (Stein and Wattenmaker 1990) provide additional evidence for noncentralized ceramic production both within Leilan and in its hinterlands. Systematic surface collections at Period IIb sites within a 15-km radius of Leilan recovered stacked fine ware bowls and wasters in at least three nearby village sites, indicating that the same fine ware bowl types were manufactured in parallel by both urban and rural craft specialists (Fig. 5). Neutron activation analyses of 178 samples of finished fine ware bowls from the village sites and from Leilan itself show clear differences in chemical composition between the urban center and its hinterland, confirming the idea of regional economic duality in craft organization (Fig. 6). There is no evidence for exchange of ceramics between the villages and the Leilan, further supporting the idea that the urban and rural sectors had low or uneven levels of regional economic integration.

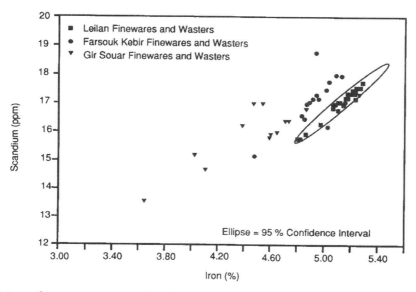

Figure 6. Neutron activation data showing compositional differences between fine ware bowls from Leilan and surrounding village production centers.

The ceramic evidence suggests that at Leilan and other city states of Mesopotamia in the 3rd millennium BC, the centralized state institutions did not try to control the production of utilitarian crafts. In any case, they could not. As a result, utilitarian crafts were produced everywhere, both in-house in the great institutions by attached specialists and outside in the other sectors of society by independent specialists. The great institutions *did* try to control the production of luxury goods or prestige goods used to validate elite status or essential in rituals: what Brumfiel and Earle (1987:5) call "politically charged commodities." The production of objects in gold, silver, and semiprecious stones was limited only to attached specialists controlled by the palace and temple sectors (Stein 1996).

What this means is that Mesopotamia had a poorly integrated, "dual" economy where each sector was essentially duplicating the functions of the other. The elites were concerned with supplying themselves with food and craft goods and in protecting their control over the prestige goods that were essential to validating their own social status. They did not or probably could not exercise this kind of economic control over the rest of Mesopotamian society.

"RESILIENT" VILLAGE PRODUCTION IN AN URBANIZED REGIONAL SYSTEM

This dual political economy also had a marked effect on agropastoral production and urban–rural relations. Mesopotamian cities and villages pursued completely different goals and economic strategies. Robert Adams, drawing on ecological models (Holling 1973), suggests that cities followed strategies of stability that aimed for consistent, high levels of productivity within very specialized, narrow ranges (Adams 1978, 1982). In other words, the cities tried to maximize production with very labor- and capital-intensive agriculture, what we would call today factory farming. Whenever possible, the cities tried to impose this pattern on the countryside as well, in order to get access to rural resources. This picture emerges clearly from zooarchaeological analyses of fourth and third millennium urban centers such as Lagash (Mudar 1982) and Malyan (Zeder 1991) and the highly centralized Ur III state (Zeder 1994).

In contrast, the villages in these urban-centered regional networks followed economic strategies of resilience. Eschewing maximal productivity as a goal, the rural sector apparently aimed for the ability to survive perturbations in a constantly changing environment. As a result, instead of pursuing a narrowly focused, specialized economy that produced large levels of surplus, Mesopotamian villages seem to have maintained diversified economies that mixed agriculture and herding. Resilient strategies do not produce as much as maximization-oriented specialized production, but they give people the flexibility of lots of options, so they can survive economically in a constantly shifting, uncertain ecological and social environment. Thus villages with generalized, resilient economic strategies would have been able to survive natural disasters such as droughts and social disasters such as the collapse of their dominant urban centers.

The resilient strategies of the rural sector conflicted with the goals of the populous cities and their centralized state institutions. As a result, we can expect to see the extent of urban control over the countryside mirrored in rural production strategies. If cities dominated the countryside, then we would expect to see the villages pulled into increasingly specialized, maximization-oriented strategies where they produced surpluses to supply to the city. However, if the villages were able to retain some kind of autonomy, then we would expect to see this reflected in their practicing strategies of resilience.

One of the best ways to reconstruct and compare the differing economic strategies of cities (and their centralized state institutions) with those of the rural sector is through the examination of pastoral production strategies, since domesticated animals (such as sheep, goats, cattle, and pigs) and their secondary products formed one of the most important classes of commodities in the state societies of the ancient Near East. Pastoral production strategies can be defined as the related set of actions and decisions by which pastoralists manipulate the composition of their herds in order to meet specific goals. Pastoralists shape herd size, species, age, and sex composition by exchanging animals, controlled breeding, and selective culling. Pastoral production strategies can be reconstructed from textual evidence, where available (e.g., Postgate 1975; Sanmartin 1993; Van De Mieroop 1993) or from zooarchaeological data in the absence of written documents (Crabtree 1990; Mudar 1982; Stein 1987; Zeder 1988, 1991). Sandford (1982, 1983) contrasts two broad classes of pastoral strategies: maximization-oriented "opportunism" and the resilient strategy of "conservatism."

An opportunistic strategy "varies the number of livestock in accordance with the current availability of forage" (Sandford 1982:62). In good years, the extra forage can be converted into capital through growth in herd size. This pastoral capital can be retained, consumed, or exchanged for easily storable forms of wealth that can be reexchanged for livestock as needed. In bad years, livestock numbers are reduced as necessary. Opportunism thus is a strategy of continual readjustment to changing range conditions based on assessments of short term risk. By contrast, a conservative strategy retains a constant number of livestock in good (i.e., rainy) and bad (dry) years alike. "A conservative strategy implies that the number of animals is not allowed to increase in the good years to utilize all the forage available, as they would then be too many for the subsequent bad years" (Sandford 1983:38).

In short, a conservative strategy sets an acceptable level of long-term risk and makes all production decisions accordingly. Conservative or risk-averse systems forego high levels of output, focusing instead on maximizing the capacity of their herds to survive and recover from serious environmental perturbations (see, e.g., Adams 1978; Sandford 1982:62). The easier it is to adjust livestock numbers upward or downward as necessary, the greater will be the tendency toward opportunistic strategies. Conversely, the more restricted the options of the herding group, the more it will tend toward conservatism.

We can reconstruct the economic strategies of the textually undocumented ancient rural sector in an urbanized state society by using excavated animal bone remains to determine whether the villagers were following specialized, maximization-oriented opportunistic strategies or generalized conservative herding strategies of resilience (Stein 1987, 1988). Specialized herding strategies would be those that tried to maximize production of a specific product like dairy products or wool, or supplying animals to the urban centers for meat. By contrast, a generalized herding strategy would just be oriented toward local subsistence and making sure that the herd is able to bounce back from possible environmental disasters like disease, harmful weather conditions, or wild predators.

Each of these different economic strategies has a different kind of optimal sheep and goat herd composition and would generate a distinctive age–sex profile of animal bone remains in the archaeological record (Payne 1973; Redding 1981). Dairy production strategies favor females and cull almost all the males at a very young age because they do not produce milk and they compete for fodder with the more valuable ewes; the resulting faunal assemblage would have large numbers (about 50% of the assemblage) of juvenile males killed between the ages of 2 and 12 months, while the remaining 50% would be almost all adult females. In wool production, the males are just as valuable as females; in fact, castrated male sheep (wethers) produce an especially fluffy wool. This specialized strategy would generate a pattern of bone remains where most of the animals would be adults, evenly divided between males and females. If the villagers' specialized production goal focused on providing animals on the hoof to a larger urban center, they would send the prime aged animals to the city, while retaining the breeding stock for themselves. This would generate a pattern where only the bones of very young and very old animals would be present in the village producing areas, while the urban consuming areas would have faunal patterns composed mainly of prime aged animals in the 2- to 3-year-old age range. In contrast to these specialized herding strategies, a generalized subsistence-oriented strategy would be a compromise among dairy and wool production and would only produce meat for local consumption. This risk averse strategy would generate a pattern of faunal remains in which all age groups are present, with males culled as they reached their optimum meat weight (at 2–3 years), while females would be culled after the age of 6 years, once their reproductive and dairy-producing capacities begin to decline (Stein 1987). The animals' ages at death can be calculated by examining patterns of tooth eruption and wear (Payne 1973). The age data let us reconstruct herd composition as a way to determine whether the rural sector was following surplus-producing specialized strategies of maximization as opposed to conservative, generalized strategies oriented toward subsistence autonomy.

Sheep and goat remains from a third millennium BC village site of Gritille in northwest Mesopotamia were examined to see if the herding economy had been drawn into an urban-dominated exchange system. Gritille is a small, 1.5-ha mound in the Euphrates river valley of southeast Turkey (Ellis and Voigt 1982; Voigt and Ellis 1981). The site is located in a well-populated and agriculturally productive area close to the Early Bronze Age urban centers of Samsat and Titris Höyük (Algaze et al. 1995; Stein 1988; Wattenmaker 1987).

Herding strategies at Gritille were calculated from a sample of 32 sheep and goat mandibles. The resulting survivorship curve was compared with expected model survivorship curves for both opportunistic and conservative herding strategies. The Gritille data were first compared with three models of opportunistic, maximization-oriented herding strategies: dairy production, wool production, and the intersite exchange of animals. The Early Bronze Age survivorship pattern did not fit any of these specialized, optimizing strategies (Fig. 7a–c). The Gritille survivorship pattern was then compared with the expected culling patterns for two conservative strategies of generalized, subsistence oriented herding: meat production for local consumption and Redding's herd security model. It is interesting to note that the model age distribution for locally oriented meat consumption closely resembles the expected survivorship curve for herd security. In other words, by following a herding strategy aimed at local meat production, villagers can at the same time very effectively minimize subsistence risk in sheep and goat husbandry. The Gritille data conform well with the expected culling patterns for both of these conservative, subsistence-oriented herding strategies (Fig. 7d).

The Early Bronze Age faunal remains suggest that village-based herders at Gritille did not try to maximize production or produce exportable surpluses of either animals or secondary products such as wool or dairy

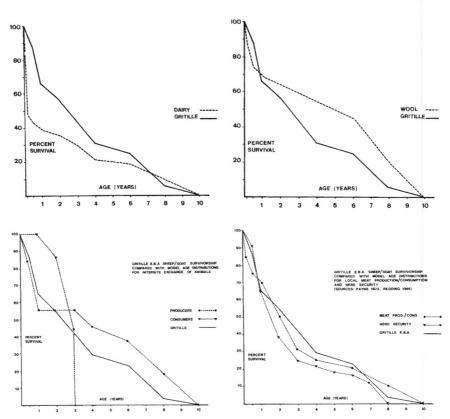

Figure 7. Sheep/goat survivorship curves for specialized and generalized pastoral production strategies compared with the evidence for animal herding at the third millennium village site of Gritille: (a) Dairy production (specialized); (b) wool production (specialized); (c) village supply of animals to urban centers (specialized); and (d) meat production for local consumption and maximization of herd security.

goods. Instead, the Gritille evidence is consistent with the kind of conservative strategies we associate with the generalized, subsistence-oriented herding that we see in rural strategies of resilience. The evidence for opportunistic, maximizing herding practices may only be present in urban sites or in a limited number of high status village households. In other words, rural production strategies for one of the most important agricultural commodities in the ancient Near East (animals and their secondary products) show a marked divergence between the economic and political strategies of the urban centers and their hinterlands. The cities appear to have attempted to impose maximizing, surplus-oriented production strategies on the areas under

SOCIAL GROUP COMPOSITION AND COMPETITION

their control in order to both feed urban populations mobilize surpluses to support the non-food-producing centralized institutions such as the temple and palace sectors.

Even in the absence of textual data, the archaeological evidence makes it clear that the villagers were practicing generalized herding strategies; they were producing for themselves and they were minimizing risk, rather than maximizing production of surplus products to supply the cities with pastoral products either as exchange items or as tribute. The fact that the villages were following resilient strategies shows that the cities could neither induce (through trade) nor compel (through taxation or tribute) the rural sector to meet urban and state demands for surplus production. The divergence between urban and rural goals, together with the evidence for a high degree of rural economic autonomy (and possibly resistance) reflect the heterogeneous organization of state societies in Mesopotamia.

CONCLUSIONS

These two case studies show the existence of important social groups pursuing economic strategies quite different from those of the temples and palaces. The significance of the relationships among these groups emerges only when we integrate archaeological information with cuneiform documents. The textual record is a priceless data source, but it is incomplete. We can only understand Mesopotamian society if we fit the textually invisible groups into a model that recognizes their importance as competing groups of social actors whose decisions and strategies affected both the organization of Mesopotamian states and their trajectories of change.

Part **V**

Conclusion

Chapter **11**

Nonlinear Paths of Political Centralization

JONATHAN HAAS

The origins and development of centralized decision making are critical issues in our understanding of how and why complex cultural systems evolve. The chapters in this volume all provide alternative perspectives on both the process and products of centralization. They also collectively attest to the combination of tinkering and pattern. In looking at the emergence of political centralization and the transition from "leaders to rulers" across times and culture areas, several major points become evident: first, there are cross-cultural commonalities in the trajectory toward increased centralization in political systems; second, societies follow many different routes in proceeding along that trajectory; and third, the "center" of centralized societies may be not be a single person—a leader or ruler who yields ultimate power and authority.

Chapter 2 in this volume by Crumley is one of those rare thought-provoking pieces that transcends theories and culture histories and provokes us to reconsider how and why cultural systems evolve. Crumley wants us to look at concepts like political centralization and the state not with a jaundiced eye, but with a fresh perspective. While a leading proponent of the concept of "heterarchies," she sees heterarchies not in opposition to hierarchies, but both as alternative organizational strategies or options that

---

JONATHAN HAAS • Department of Anthropology, The Field Museum, Chicago, Illinois 60605.

From Leaders to Rulers, edited by Jonathan Haas. Kluwer Academic/Plenum Publishers, New York, 2001.

provide different advantages and disadvantages. A society may be heterarchically organized at one time and hierarchically at another. Indeed, heterarchical organization can exist alongside hierarchical organization in the same political system—a principle that is quite consistent with Stein's analysis of alternative power centers coexisting in a political centralized system.

Crumley also wants us to look at the evolution of complex political systems in nonlinear terms and develop models that more effectively incorporate the kinds of political complexity expressed in the preceding chapters. She is pushing cultural evolution to more fully integrate the agency of individuals and proposes that the evolution of ancient states can and should be analyzed as products of self-organization. [In terms of the intellectual history of cultural evolution, it is noteworthy that Crumley reaches back to Service (1971a) for insights on the role of local level associations in the evolution of state societies.] A "bottom-up" approach to the evolution of complex polities, dependent on the agency of actors on the cultural landscape, is quite different from the traditional "top-down" models that have tended to predominate in the archaeological literature. Crumley ultimately leaves us with a challenge to creatively rethink the emergence and development of politically centralized polities.

In Creamer's chapter on the Pueblos of the American Southwest, she provides an analysis of the alternative political and economic strategies applied by diverse communities within a single region. This is a world area that appears as somewhat of an enigma in cross-cultural considerations of political evolution. At the time of European contact in the sixteenth century, Puebloan societies were strong and economically prosperous. Population size and density were both relatively high (with many large villages occupied at the same time) and significant food surpluses were being produced in all the villages. Cultural, ethnic, and linguistic boundaries separated groups of villages from each other and interpueblo warfare was common if not endemic. Basically, all the elements conducive to the evolution of politically centralized systems seem to be present in the Puebloan region, but centralized polities never fully developed on the Pueblo landscape.

What Creamer's analysis shows is that the Pueblo communities were experimenting with diverse options in their efforts to compete effectively with the neighbors and produce adequate resources—both quantity and variety—for their local populations. They were in an active period of tinkering with social, economic, and political organization. Although some pueblos appear to have been largely independent, others were more tightly integrated into subregional alliances and confederacies. There were indications of economic specialization in some areas and not in others, significant exchange systems among some communities and little exchange in others.

NONLINEAR PATHS OF POLITICAL CENTRALIZATION 237

While certain religious practices (e.g., Kachinas) were common to almost all the villages, their material manifestations in architecture and site layout differed from one village/area to the next. The picture painted by Creamer is one of a region in transition. Although full political centralization around the office of a chief had not emerged in the Pueblo region by the time of the arrival of the Spaniards, there were certainly central village councils and war leaders. In this manifestation, she argues the region has the attributes of corporate as opposed to network organization (see Chapter 8, this volume). Her observations about eventual centralization of the Puebloan system after European contact during the Pueblo Revolt of 1680 offer a tantalizing hint of the potential evolutionary trajectory of the system had it not been adumbrated by the arrival of Spanish missionaries and soldiers.

In Chapter 4, Gilman provides an important reassessment of the development of political systems in southeastern Spain during the late Neolithic and the Copper and Bronze ages. For those following the specifics of archaeology in the region, this chapter offers a significant new interpretation of the data. Gilman reconsiders the nature and scope of centralized society in southeastern Spain and the role of leaders in that society. Since the work of Gilman and others in this region has played a substantial role in furthering our understanding of the relationship between agriculture and the power of emergent leaders, his current analysis has broad bearing on our understanding of the initial steps in the cross-cultural trajectory of political centralization.

Elsewhere, Gilman (1981, 1991) has presented a cogent case that the development of labor-intensive orchards provides an avenue for the exercise of power by emergent leaders in southeastern Spain. Although he does not discount this developmental model in the current analysis, he does argue that without additional bases of power derived through trade, specialized production, or other activities, the power of emergent leaders was limited and fragile. The social stratification and incipient centralization that emerged in the Copper Age (3500–2250 BC) was not solidified and institutionalized in the ensuing Bronze Age (2250–1500 BC). This case from southeastern Spain offers an excellent illustration that the trajectory toward political centralization is not inexorably linear and unidirectional.

Although politically centralized polities may be economically efficient and culturally competitive, there also is an inherent tension in centralized systems between the institutions of leadership (see Chapter 5, this volume) and the independent decision-making authority of the component parts of the system (see Chapter 10, this volume). Although the evidence from many world areas shows that leadership is a stubborn institution once it emerges, Gilman shows that the tension is not always resolved in favor of increasing centralization and consolidation of the power of leaders.

In Chapter 5, Kristiansen adds a further perspective on how leadership comes to be institutionalized on the local level using the early chiefdoms of Bronze Age Europe as his archaeological laboratory. In this case, he argues that in the early stages of development, two complementary leadership roles emerged—one as a war chief and the other as a chiefly priest. This duality of roles is represented both symbolically in the form of twin figures and in burial contexts with different kinds of associated funerary objects.

Continuing on from the discussion in Kristiansen, the Maya case—with a rich body of archaeological data and inscriptions—sheds further light on the role of ideology in the institutionalization of rulership and the interrelationship between ideological leadership and political authority. McAnany, in Chapter 7, places the Maya in a cross-cultural framework as she looks at the rise of political authority. While much of the archaeological discussion of the development of political centralization has emphasized the role of economics and warfare, McAnany highlights the important complementary role of ideology. She sees ideology as providing a critical vehicle for the initial emergence of leadership and then the consolidation of rulership in evolving polities. While economic production, exchange, and warfare were the material base of Mayan society, the rulers gained the "mandate to rule" through an ideological connection to their ancestors and to the cosmos.

McAnany's analysis of political centralization in the Mayan area of Mesoamerica also focuses on the fine mechanics of evolution. She brings her analytical lens down to the level of the individual, not in a historical sense but in a contextual one. She presents a cogent interpretation of how the role of shaman provides an avenue for the early appearance of leadership in Mayan society and then in turn how the position of spiritual leadership provides the means for the transformation of shaman leaders into the rulers of Mayan polities. She brings the analysis down to the level of the individual and reinforces the point that cultures evolve through the actions of individuals, not through the mindless actions of some monolithic cultural whole. It is interesting that while McAnany specifically considers the role of the individual—where we might least expect to find cross-cultural pattern and process—she also shows that the paths followed by the Mayan rulers were not at all dissimilar to the paths followed by evolving rulers in other world areas. This is yet another example of how human agency and tinkering play clear roles in the evolution of cultural systems and how consistent patterns evolve out of that agency and tinkering across time and space.

In Chapter 8, Feinman also considers the role of individual Mayan rulers and compares them with the relatively faceless rulers found at contemporary Teotihuacan in central Mexico. In this comparison, he highlights fundamental differences in the nature of rulership found in the two polities: a network strategy for the Maya and a corporate strategy for Teotihuacan.

NONLINEAR PATHS OF POLITICAL CENTRALIZATION

The Maya, where the deeds and stature of individual rulers is abundantly manifested in art, architecture, and hieroglyphs, are a fully centralized and hierarchical polity—the network model in Feinman's analysis. In contrast, at Teotihuacan, where the manifestations of individual, recognizable rulers are largely absent, political centralization and social hierarchy are only partially realized—Feinman's corporate model. On a broad cross-cultural and evolutionary scale, both Mayan and Teotihuacan polities have a highly centralized decision making apparatus, in the sense that a centralized governmental body is making and implementing decisions that apply to the polity as a whole. In both cases, the state government exercises considerable power over its respective populations and directs the behaviors of those populations in obvious and diverse ways. However, the differences in political strategies between the two polities reflect significant differences in the role, nature, and function of government, and these differences in turn have important ramifications for our understanding of the evolution of political centralization.

The intriguing question that Feinman raises at the end of his contribution is why network strategies arise under some conditions while corporate strategies arise under others. This is a particularly interesting question for the Maya, since this is not one polity but many independent (though interacting) polities, all characterized by the network strategy. While this pattern would have emerged through a process of evolutionary tinkering (as can be seen in McAnany's analysis), the different Mayan polities all ultimately developed a remarkably similar kind of network strategy. It would appear in this case that in a competitive cultural environment, the emergence of a network-based strategy in one polity plays a strong role in convergence on that same strategy by all the polities in the interaction sphere (see Price 1977; Culbert 1991). In contrast, while Teotihuacan certainly had competitors outside the valley of Mexico, it was clearly the overwhelmingly dominant polity within the valley for more than half a millennium. It would be interesting to look at political strategies in the Valley of Mexico at an earlier time period when Teotihuacan had more immediate competitors.

The theme of alternate political strategies is further expanded in Chapter 6, where Earle compares economic and political organization in centralized chiefdom societies from three different world areas: Denmark, Hawaii, and the Andean highlands of Peru. Earle has written extensively about these three case studies, and the current chapter combines the notions of network versus corporate political strategies with the concepts of wealth versus staple finance to explain patterns of variability found in different chiefdoms. Of all the chapters in the volume, Earle's provides the clearest picture of the potential intersection of multiple factors in the evolution of complex centralized political systems. He successfully "decouples" the process of political centralization from strict pyramidal hierarchies and from the social and economic self-aggrandizement of leaders. In other words, centralized systems can exist without readily identifiable individual decision makers and without individuals accumulating conspicuous personal power.

In this innovative expansion of his previous work on chiefdoms, Earle also uses the three culture areas to show how aspects of the built landscape serve as physical indicators of the organizational structure of divergent cultural systems. While all three cases are at a similar chiefdom level, they have profoundly different physical manifestations on the ground. Earle then uses these landscape qualities as both windows into and reflections of differences in the nature of sociopolitical organization and the role of leaders in the three areas. Extrapolating from Earle's analysis, it becomes an interesting question to compare the widely divergent built landscapes in even more centralized state-level political systems to look for insights into structural differences. Thus, in the early states of ancient China, for example, there is a general absence of large, aboveground monumental architecture but enormous belowground tombs of royalty. In contrast, along the coast of Peru the earliest emergent polities all constructed huge platform mounds, but there is a general absence of mortuary architecture associated with the rulers. While never intended to generically address such historical specifics, Earle's analysis provides a good starting point for examining patterned differences in widely separated cultural systems.

In Chapter 9, Billman shows how the characteristics of the built landscape can be effectively used to extract further inferences about the exercise of power by rulers in centralized societies. In this unique environment of dry coastal valleys transected by rivers coming out of the Andes, huge structures of adobe and stone extend back through more than 4000 years of Peruvian prehistory. Billman uses these very public mound structures as a measure to determine the emergence and relative strength of power holding elite rulers. By plotting the distribution and chronology of these mounds, he has been able to identify the central coast as the location of the earliest appearance of state-level societies with powerful rulers. (It is interesting to note that these early platform mounds do not seem to be accompanied by equally early high status architecture, and thus in Feinman's model would tend toward the corporate end of the spectrum of political strategies.)

With monumental constructions as indicators of early political centralization, Billman then looks at the central coast and asks why this area in particular seems to be the locus of emergent polities. To find an answer, he looks at regional environmental patterns and finds that the central coast is at the nexus of two broad trends. The first trend is that coastal valleys in Peru get smaller as one moves from north to south, that is, northern valleys tend to have more arable land and southern valleys have less. The second trend relates to the occurrence of El Niño events. In Peru, El Niños are manifested by the extension of warm equatorial waters down the normally cold Peruvian coastline. As a result, during the El Niño years there is heavy rainfall on the normally dry coast and the cold-water-dependent marine biota disappears from the shoreline. The combined consequences are often devastating for coastal populations dependent on irrigation farming of the desert valleys and on fish and shellfish from the coast. While the consequences of El Niños are relatively consistent from one part of the coast to another, the frequency of occurrences are not. The coastal valleys further north and closer to the equator experience more frequent El Niños; those to the south experience them less often. The central coast, then, lies at an intermediate point in terms of valley size and frequency of El Niño events. It is this intermediate position, Billman argues, that is conducive to the emergence of power-holding elites and the formation of very early centralized polities.

Chapter 10 by Stein is groundbreaking in looking at the complex power-holding relationship found in what is generally perceived to be a highly centralized, strictly hierarchical state society—ancient Mesopotamia. What Stein demonstrates empirically is that the concept of political centralization has to be considered in relative terms. Here he is reinforcing a similar point made in several other chapters (e.g., Chapters 2, 5, and 8) that centralized societies may not have a single omnipotent center and indeed may not be unicentric. In the case of Mesopotamia, Stein shows that geographically and economically peripheral components of Mesopotamian polities could be quite autonomous of the centralized governing bodies of those polities. These peripheral elements effectively constitute alternative power centers, both complementing and possibly competing with the ruling elites of Mesopotamian society.

While this analysis exposes the complexities of the seemingly simple concept of "centralization," it enriches rather than complicates our understanding of the evolution of politically centralized societies. Mesopotamian societies are not decentralized in any meaningful sense, nor are they heterarchies in the sense of Crumley nor corporate strategies in the sense of Feinman. They can best be described as centralized, hierarchical, network-type societies with complex arrangements in the distribution and exercise of power. Stein thus brings in yet another dimension that must be considered in studying the evolutionary trajectory of political centralization.

Intuitively his analysis makes a great deal of sense if we just consider our own contemporary political systems. There is an enormously complex relationship between all state governments and an array of regional authorities, rebel organizations, business interests, and "criminal elements." Why would prehistoric polities be any different? In interesting ways, Stein's analysis shows that the maturation of evolutionary models is an ongoing process. The more we learn about the emergence and subsequent development of politically centralized models, the more productive our analyses and the richer our understanding of cultural evolution become.

Individually and collectively the chapters in this volume illustrate the complex evolution of political centralization. Clearly as a cross-cultural process, centralization is not unilinear in the sense of an inexorable trend toward greater and greater concentration of power in the hands of fewer and fewer people. Societies in different places and at different times pursue various political strategies that fall along multiple organizational spectra: corporate versus network; heterarchy versus hierarchy; wealth versus staple finance; coercive versus benificent government; powerful versus weak rulers. Emergent and developing centralized polities have the same kind of variations that are witnessed in the modern world of fully developed mature states.

Despite of the diversity manifested in diverse archaeological sequences, the chapters also exemplify the general cross-cultural trajectory toward increasing political centralization in the broadest sense—the few making decisions for the many. The broad path of this trajectory follows from nascent beginnings in the experimental political strategies of the Rio Grande pueblos through the combination of leadership arrangements found in the chiefdoms of Europe and Polynesia to the rule of kings and queens in the powerful states of Mesoamerica, Peru, and Mesopotamia. While the histories of each society are unique, there are common patterns that crosscut cultural lines and allow us to extract process from history.

Placed in the context of the history of evolutionary studies, the chapters in this volume build on the foundations laid by their predecessors. While the various stage models can be justly criticized for "coupling" together all the components of evolving cultural systems, there is nevertheless a general recognition that cultural evolution is marked by consistent cross-cultural landmarks. The transition from nomadic hunters and gatherers to the nation states of the modern world is not one of gradual accretionary change but of a series of qualitative evolutionary transformations. The initial emergence of specialized leadership and the shift from leaders to rulers are then integral components of this transformational process. In interesting ways, by analytically "decoupling" political organization from other aspects of evolving cultural systems, we have the freedom to "recouple" politics with economics, religion, warfare, demographics, ans so forth, in innovative alternative combinations. The purpose of such recouplings is not to build yet another set of typological stages, but to provide fresh insights into how and why cultures evolve over time and why certain kinds of cultural configurations recur consistently in the global record of humanity.

NONLINEAR PATHS OF POLITICAL CENTRALIZATION

Studying the process of political centralization through the tools and data of archaeology provides insights into human social and political organization that extend from the ancient past into the contemporary world. Centralized polities dominate the global landscape today, with a spectrum of organizational strategies from hierarchical dictatorships to the heterarchy of the United Nations Security Council. The corporate order of presidents and prime ministers coexists alongside the faceless network bureaucracy of the European Union and OPEC. The archaeological record of past societies gives a diachronic perspective on how and why different kinds of systems evolve and over time collapse. The chapters in this volume are both a product of and divergence from the collective history of evolutionary studies and anthropology and at the same time have an important place in the social science dialogue over the role and future of political organization in a changing, evolving world.

References

Abrams, E., 1994, *How the Maya Build Their World*. University of Texas Press, Austin.
Adams, E. C., 1989, The Homol'ovi Research Program. *Kiva* 54: 175–194.
Adams, E. C., 1991, *The Origins and Development of the Pueblo Katsina Cult*. University of Arizona Press, Tucson.
Adams, R. N., 1975, *Energy and Structure: A Theory of Social Power*. University of Texas Press, Austin.
Adams, R. McC., 1966, *The Evolution of Urban Society*. University of Chicago Press, Chicago.
Adams, R. McC., 1972, Patterns of Urbanization in Early Southern Mesopotamia. In *Man, Settlement, and Urbanism*, edited by P. Ucko, R. Tringham, and G. W. Dimbleby, pp. 735–749. Duckworth, London
Adams, R. McC., 1974, The Mesopotamian Social Landscape: A View from the Frontier. *Bulletin of the American Schools of Oriental Research* (Suppl.): 1–20.
Adams, R. McC., 1978, Strategies of Maximization, Stability, and Resilience. *Proceedings of the American Philosophical Society* 122(5): 329–335.
Adams, R. McC., 1981, *Heartland of Cities*. University of Chicago Press, Chicago.
Adams, R. McC., 1982, Property Rights and Functional tenure in Mesopotamian Rural communities. In *Societies and Languages of the Ancient Near East: Studies in Honour of I.M. Diakonoff*, (edited by) N. Postgate, pp. 1–14. Aris and Phillips, Warminster, England.
Adams, R. McC., and Nissen, H. 1972, *The Uruk Countryside*. University of Chicago Press, Chicago.
Adler, M. (ed.), 1996, *The Pueblo III Period in the Northern Southwest*. University of Arizona Press, Tucson.
Algaze, G., 1989, A New Frontier: First Results of the Tigris–Euphrates Archaeological Reconnaissance Project. 1988. *Journal of Near Eastern Studies* 48(4): 241–281.
Algaze, G. (ed.), 1990, *Town and Country in Southeastern Anatolia, Volume 2: The Stratigraphic Sequence at Kurban Höyük*. University of Chicago, Oriental Institute Publication 110, Chicago.
Algaze, G., Breuninger, R., Lightfoot, C., and Rosenberg, M., 1991, The Tigris–Euphrates Archaeological Reconnaissance Project: A Preliminary Report of the 1989–1990 Seasons. *Anatolica* 17: 175–240.

REFERENCES

Algaze, G., Kelly, J., Matney, T., and Schlee, D., 1995, Titris Höyük: A Small Early Bronze Age Urban Center in Southeast Anatolia: The 1994 Season. *Anatolica* 21: 13–64.

Almagro, M., 1962, El Ajuar del "Dolmen de la Pastora" de Valentina del Alcor (Sevilla): Sus Paralelos y su Cronología. In *Trabajos de Prehistoria*, volume 5, Instituto Español de Prehistoria, Madrid.

Almagro, M., and Arribas, A., 1963, El Poblado y la Necrópolis Megalíticos de Los Millares (Santa Fe de Mondújar, Almería). *Bibliotheca Praehistorica Hispana*, vol. 3. Consejo Superior de Investigaciones Científicas, Madrid.

Almagro Gorbea, M. J., 1973, El poblado y la necrópolis de El Barranquete (Almería). *Acta Arqueológica Hispánica*, vol. 6. Ministerio de Educación y Ciencia, Madrid.

Alva, W., 1986, *Las Salinas de Chao: Un Asentamiento Temprano, Obervaciones y Problematica*. Kommission fur Allgemeine und Vergleichende Archaolige des Deutschen Archaologischen Institus Bonn, Band 34, Munchen.

Ames, K. M., 1991, Sedentism: A Temporal Shift or a Transitional Change in Hunter-Gatherer Mobility Patterns. In *Between Bands and States*, edited by S. A. Gregg, pp. 108–134, Occasional Papers No. 9, Center for Archaeological Investigations, Southern Illinois University at Carbondale.

Anderla, G., Dunning, A., and Forge, S., 1997, *Chaotics: An Agenda for Business and Society in the 21st Century*. Praeger, Westport.

Anderson, B. R., 1972, The Idea of Power in Javanese Culture. In *Culture and Politics in Indonesia*, edited by C. Holt, pp. 1–70, Cornell University Press, Ithaca, New York.

Andrews, V. E. W., 1990, The Early Ceramic History of the Lowland Maya. In *Vision and Revision in Maya Studies*, edited by F. S. Clancy and P. D. Harrison, pp. 1–19, University of New Mexico Press, Albuquerque.

Aner, E., and Kersten, K., 1978, Die Funde der älteren Bronzezeit des nordischen Kreises in *Dänemark, Schleswig-Holstein und Niedersachsen*. Band IV, Südschleswig-Ost. Verlag Nationalmuseum København, Karl Wachholtz, Neumünster.

Aner, E., and Kersten, K., 1984, Nordslesvig-Nord: Haderslev Amt. In *Die Funde der älteren Bronzezeit des nordisches Kreises in Dänemark, Schleswig-Holstein und Niedersachsen*, vol. 7. Karl Wachholtz Verlag, Neumunster.

Angelini, M. L., 1997, *Crafting Pottery: A Study of Formative Maya Ceramic Technology at K'axob, Belize*. PhD dissertation, Department of Archaeology, Boston University, University Microfilms International, Ann Arbor.

Aperlo, P., 1994, *Crude but Effective. Stone Tools and Household Production in Early Bronze Age Thy, Denmark*. Master's Thesis, Archaeology Program, University of California at Los Angeles.

Araus, J. L., Febrero, A., Rodríguez-Ariza, M. O., Molina, F., Camalich, M. D., Martín, D., & Volta, J., 1997, Identification of Ancient Irrigation Practices Based on the Carbon Isotope Discrimination of Plant Seeds: A Case Study from the Southeast Iberian Peninsula. *Journal of Archaeological Science*, 24: pp. 729–740.

Archi, A., 1985, The Royal Archives of Ebla. In *Ebla to Damascus*, edited by H. Weiss, pp. 140–148. Smithsonian Institution Press, Washington, DC.

Arnold, J. E., 1991, Transformations of a Regional Economy. *Antiquity* 65(249): 953–962.

Arribas Palau, A., 1986, La época del Cobre en Andalucía Oriental: Perspectivas de la Investigación Actual. In *Actas del Congreso "Homenaje a Luis Siret" (1934–1984)*, edited by Oswaldo Arteaga, pp. 159–196. Consejería de Cultura de la Junta de Andalucía, Sevilla.

Arribas, A., & Molina, F., 1979, *El poblado de 'Los Castillejos' en las Peñas de los Gitanos (Montefrío, Granada): Campaña de Excavaciones de 1971, el Corte n° 1*. Universidad de Granada, Granada.

REFERENCES

Arribas, A., & Molina, F., 1982, Los Millares: Neue Ausgrabungen in der kupferzeitlichen Siedlung (1978–1981). *Madrider Mitteilungen* 23: 9–32.
Arribas, A., Molina, F., Sáez, L., de la Torre, F., Aguayo, P., Bravo, A., & Suárez, A., 1983, Excavaciones en Los Millares (Santa Fe de Mondújar, Almería): Campañas de 1982 y 1983. *Cuadernos de Prehistoria de la Universidad de Granada* 8: 123–147.
Arteaga, O., 1992, Tribalización, Jerarquización y Estado en el Territorio de El Argar. *Spal* 1: 179–208.
Ayala Juan, M. M., 1991, *El Poblamiento Argárico en Lorca: Estado de la Cuestión*. Real Academia Alfonso X el Sabio, Murcia.
Bailey, F. G. (ed.), 1971, *Gifts and Poison: The Politics of Reputation*. Basil Blackwell, Oxford.
Baines, J., and Yoffee, N., 1998, Order, Legitimacy, and Wealth in Ancient Egypt and Mesopotamia. In *Archaic States*, edited by G. Feinman and J. Marcus, pp. 199–260. School of American Research Press, Santa Fe.
Barnett, H., 1953, *Innovation: The Basis of Cultural Change*. McGraw-Hill, New York.
Barrera Morate, J. L., Martínez Navarrete, M. I., San Nicolás del Toro, M., and Vicent García, J. M., 1987, El Instrumental Lítico Pulimentado de la Comarca Noroeste de Murcia: Algunas Implicaciones Socio-económicas del Estudio Estadístico de su Petrología y Morfología. *Trabajos de Prehistoria*, 47: pp. 87–146.
Bartlett, R. V., and Baber, W. F., 1987, Matrix Organization Theory and Environmental Impact Analysis: A Fertile Union? *Natural Resources Journal* 27: 605–615.
Bartlett, M. L., and McAnany, P. A., 2000, "Crafting" Communities: The Materialization of Formative Maya Identiities. In *The Archaeology of Communities: A New World Perspective*, edited by M. A. Canato and J. Yaeger, pp. 102–122. Routledge Press, London.
Barton, C. M., Rubio Gomis, F., Miksicek, C. A., and Donahue, D. J., 1990, Domestic Olive. *Nature* 346: 518–519.
Bateson, G., 1972, *Steps Toward an Ecology of Mind*. Ballentine, New York.
Baugh, T. G., and Nelson, Jr., F. W., 1987, New Mexico Obsidian Sources and Exchange on the Southern Plains. *Journal of Field Archaeology* 14: 313–329.
Bayer, L., 1994, *Santa Ana: The People, the Pueblo, and the History of Tamaya*. University of New Mexico Press, Albuquerque.
Benedetto, R. F., 1985, *Matrix Management: Theory in Practice*. Kendall/Hunt Publishing Co, Dubuque.
Benedict, R., 1934, *Patterns of Culture*. Houghton Mifflin, New York.
Berlin, H., 1958, El Glifo "Emblema" en las Inscripciones Mayas. *Journal de la Société des Américanistes* 47: 111–119.
Bettinger, R. L., 1991, *Hunter-Gatherers: Archaeological and Evolutionary Theory*. Plenum Press, New York.
Billman, B., 1996, *The Evolution of Prehistoric Political Organizations in the Moche Valley, Peru*. Unpublished Ph.D. dissertation, Department of Anthropology, University of California, Santa Barbara, California.
Billman, B., 1997, Population Pressure and the Origins of Warfare in the Moche Valley, Peru. In *Integrating Archaeological Demography: Multidisciplinary Approaches to Prehistoric Populations*, edited by R. R. Paine, pp. 285–310. Center for Archaeological Investigations, Occasional Paper 24, Carbondale.
Billman, B., 1999, Reconstruction Cycles of Political Power: Irrigation, Social Stratification, Warfare, and the Origins of the Moche State. In *Settlement Pattern Studies in the Americas: Fifty Years Since Viru*, edited by B. R. Billman and G. M. Feinman, pp. 131–159. Smithsonian Institution Press, Washington, DC.

REFERENCES

Binford, L., 1964, A Consideration of Archaeological Research Design. *American Antiquity* 29: 425–441.

Bird, J. B., 1994, Prehistoric Farmers in the Chicama and Viru. In *A Reappraisal of Peruvian Archaeology*, edited by W. C. Bennett, pp. 1–65. Society for American Archaeology Memoir No. 4, Salt Lake City.

Blackman, M. J., Stein, G. J., and Vandiver, P., 1993, The Standardization Hypothesis and Ceramic Mass Production: Compositional, Technological, and Metric Indices of Craft Specialization at Tell Leilan (Syria). *American Antiquity* 58(1): 60–80.

Blake, M., Clark, J. E., Voorhies, B., Michaels, G., Love, M. W., Pye, M. E., Demarest, A. A., and Arroyo, B., 1995, Radiocarbon Chronology for the Late Archaic and Formative Periods on the Pacific Coast of Southern Mesoamerica. *Ancient Mesoamerica* 6: 161–183.

Blanton, R. E., 1983, The Founding of Monte Alban. In *The Cloud People: Divergent Evolution of the Zapotec and Mixtec Civilizations*, edited by K. V. Flannery and J. Marcus, pp. 83–87. Academic Press, New York.

Blanton, R. E., 1998, Beyond Centralization: Steps toward a Theory of Egalitarian Behavior in Archaic States. In *Archaic States*, edited by G. M. Feinman and J. Marcus, pp. 135–172. School of American Research Press, Santa Fe.

Blanton, R. E., Kowalewski, S. A., Feinman, G. M., and Finsten, L. M., 1993, *Ancient Mesoamerica: A Comparison of Change in Three Regions*, 2nd edition. Cambridge University Press, Cambridge.

Blanton, R. E., Feinman, G. M., Kowalewski, S. A., and Peregrine, P. N., 1996, A Dual-Processual Theory for the Evolution of Mesoamerican Civilization. *Current Anthropology* 37: 1–14.

Bloch, M., 1971, *Placing the Dead: Tombs, Ancestral Villages, and Kinship Organization in Madagascar*. Seminar Press, London.

Boas, F., 1911, *The Mind of Primitive Man*. Macmillan, New York.

Boehm, C., 1996, Emergency Decisions, Cultural-selection Mechanics, and Group Selection. *Current Anthropology* 37(5): 763–793.

Boehm, C., 1997, Impact of the Human Egalitarian Syndrome on Darwinian Selection Mechanics. *The American Naturalist* 150:S100–S121.

Boone, E. H., 1994, *The Aztec World*. St. Remy Press and Smithsonian Books, Montreal and Washington, DC.

Bosch Gimpera, P., 1932, *Etnología de la Península Ibérica*. Alpha, Barcelona.

Bourdieu, P., 1977, *Outline of a Theory and Practice*. Cambridge University Press, Cambridge.

Bourdieu, P., 1979, *Algeria in 1960*, Cambridge University Press, Cambridge.

Boyd, R. and Richerson, P. J., 1985, *Culture and Evolutionary Process*. University of Chicago Press, Chicago, 1985.

Bradley, R., 1996, Sacred Geography. *World Archaeology* 28 (2).

Braidwood, R. J. and Willey, G. R. (eds.), 1962, *Courses toward Urban Life*. Aldine, Chicago.

Brandt, E. A., 1977, The role of Secrecy in a Pueblo Society. In *Flowers of the Wind: Papers on Ritual, Myth and Symbolism in California and the Southwest*, edited by Thomas C. Blackburn, pp. 11–28. Ballena Press, Socorro.

Brandt, E. A., 1994, Egalitarianism, Hierarchy, and Centralization in the Pueblos. In *The Ancient Southwestern Community*, edited by W. H. Wills and R. D. Leonard, pp. 9-23. University of New Mexico Press, Albuquerque.

Braun, D. P., 1990, Selection and Evolution in Nonhierarchical Organization. In *The Evolution of Political Systems: Sociopolitics in Small-Scale Sedentary Societies*, edited by S. Upham, pp. 62–86. Cambridge University Press, Cambridge.

Braun, D. P., 1995, Style, Selection, and Historicity. In *Style, Society, and Person: Archaeologi-*

cal and Ethnological Perspectives, edited by C. Carr and J. E. Neitzel, pp. 124–141. Plenum Press, New York.

Brinkman, J. A., 1984, Settlement Surveys and Documentary Evidence: Regional Variation and Secular Trend in Mesopotamian Demography. *Journal of Near Eastern Studies* 43(3): 169–180.

Broholm, H. C. and Hald, M., 1940, *Costumes of the Bronze Age in Denmark*. Nyt nordisk Forlag, Copenhagen.

Brown, J. A., 1997, The Archaeology of Ancient Religion in the Eastern Woodlands. In *Annual Review of Anthropology* 26: 465–485.

Brumfiel, E., 1976, Regional growth in the eastern Valley of Mexico: A Test of the "Population Pressure" Hypothesis. In *The Early Mesoamerican Village*, edited by K. V. Flannery, pp. 234–249. Academic Press, New York.

Brumfiel, E., 1992, Distinguished Lecture in Archaeology: Breaking and Entering the Ecosystem—Gender, Class, and Faction Steal the Show. *American Anthropologist* 94(3): 551–567.

Brumfiel, E., 1994, Introduction. In *The Economic Anthropology of the State*, edited by E. Brumfiel, pp. 1–16. University Press of America, Lanham, MD.

Brumfiel, E., and Earle, T., 1987, Specialization, Exchange, and Complex Societies: An Introduction. In *Specialization, Exchange, and Complex Societies*, edited by E. Brumfiel and T. Earle, pp. 1–9. Cambridge University Press, Cambridge

Bryson, R. A., and Murray, T. J., 1977, *Climates of Hunger*. University of Wisconsin Press, Madison.

Buikstra, J., Castro, P., Chapman, R., González Marcén, P., Hoshower, L., Lull, V., Micó, R., Ruiz, M., & Encarna Sanahuja Yll, M., 1995, Approaches to Class Inequalities in the Later Prehistory of South-east Iberia: The Gatas project. In *The Origins of Complex Societies in Late Prehistoric Iberia*, edited by K. T. Lillios, pp. 153–168. International Monographs in Prehistory, Ann Arbor.

Burger, R. L., 199, *Chavin and the Origins of Andean Civilization*. Thames and Hudson, New York.

Burjachs, F., and Riera, S., 1996, Canvis Vegetals i Climàtics Durant el Neolític a la Façada Mediterrània Ibèrica. In *Rubricatum (Actes, I Congrés del Neolític a la Península Ibèrica, Formació i implantació de les comunitats agrícoles, Gavà-Bellaterra, 27, 28 i 29 de març de 1995)*, volume 1, pp. 21–27. Museu de Gavà, Gavà.

Burke, P., 1990, *The French Historical Revolution: The Annales School 1929–1989*. Stanford University Press, Stanford.

Business International Corporation, 1981, *New Directions in Multinational Corporate Organization*. Business International Corporation, New York.

Cabrera Castro, R., Sugiyama, S., and Cowgill, C. L., 1991, The Templo de Quetzalcoatl Project at Teotihuacan. *Ancient Mesoamerica* 2: 77–92.

Çambel, A. B., 1993, *Applied Chaos Theory: A Paradigm for Complexity*. Academic Press, New York.

Cameron, M. A. S., 1967, Unpublished Fresco Fragment at a Chariot Composition from Knossos. *Archaeologica Atlantica*.

Cancian, F., 1965, *Economics and Prestige in a Maya Community: The Religious Cargo System in Zinacantan*. Stanford University Press, Stanford.

Capel, J., Reyes Delgado, E. A., Nuñez, T., and Molina, F., 1998, Palaeoclimatic Identification Based on an Isotope Study of Travertine from the Copper Age Site at Los Millares, Southeastern Spain. *Archaeometry* 40: 177–185.

Carneiro, R. L., 1970, A Theory of the Origins of the State. *Science*, 169: 733–738.

Carneiro, R. L., 1972, From Autonomous Villages to the State, a Numerical Estimation. In

Population Growth: Anthropological Implications, edited by B. Spooner, pp. 64–77. MIT Press, Cambridge.
Carneiro, R. L., 1981, The Chiefdom: Precursor of the State. In T*he Transition to Statehood in the New World*, edited by G. D. Jones and R. R. Kautz, pp. 37–79. Cambridge University Press, Cambridge.
Carrasco, P., 1971, Social Organization of Ancient Mexico. In *Handbook of Middle American Indians, Volume 10, Archaeology of Northern Mesoamerica, Part One*, edited by G. F. Ekholm and I. Bernal, pp. 349–375. University of Texas Press, Austin.
Castro, P. V., Chapman, R. W., Gili, S., Lull, V., Micó, R., Rihuete, C., Risch, R., and Sanahuja, M. E., 1999, Agricultural Production and Social Change in the Bronze Age of Southeast Spain: The Gatas Project. *Antiquity* 73: 846–856.
Castro, P. V., Gili, S., Lull, V., Micó, R., Rihuete, C., Risch, R., and Sanahuja Yll, M. E., 1998, Teoría de la Producción de la Vida Social: Mecanismos de Explotación en el Sudeste Ibérico. *Boletín de Antropología Americana* 33: 25–77.
Cavalli-Sforza, L. L. and Feldman, M. W., 1981, *Cultural Transmission and Evolution*. Princeton University Press, Princeton.
Chagnon, N., 1983, *Yanomamí: The Fierce People,* 3rd edition. Holt, Rinehart, and Winston, New York.
Chang, K. C., 1983, *Art, Myth, and Ritual: The Path to Political Authority in Ancient China*. Harvard University Press, Cambridge.
Chapman, R. W., 1975, *Economy and Society Within Later Prehistoric Iberia: A New Framework*. Ph.D. dissertation, University of Cambridge, Cambridge.
Chapman, R. W., 1978, The Evidence for Prehistoric Water Control in South-east Spain. *Journal of Arid Environments* 1: 261–274.
Chapman, R. W., 1981, Archaeological Theory and Communal Burial in Prehistoric Europe. In *Pattern of the Past: Studies in Honour of David Clarke*, edited by I. Hodder, G. Isaac, & N. Hammond, pp. 387–411. Cambridge University Press, Cambridge.
Chapman, R. W., 1982, Autonomy, ranking, and resources in Iberian prehistory. In *Ranking, Resource and Exchange: Aspects of the Archaeology of Early European Society*, edited by C. Renfrew & S. Shennan, pp. 46–51. Cambridge University Press, Cambridge.
Chapman, R. W., 1984, Early Metallurgy in Iberia and the Western Mediterranean: Innovation, Adoption and Production. In *The Deya Conference of Prehistory: Early Settlement in the Western Mediterranean Islands and Their Peripheral Areas*, edited by W. H. Waldren, R. Chapman, J. Lewthwaite, and R-C. Kennard, pp. 1139–1165. BAR International Series 229, Oxford.
Chapman, R. W., 1990, *Emerging Complexity: The Later Prehistory of South-East Spain, Iberia and the West Mediterranean*. Cambridge University Press, Cambridge.
Charvat, P., 1982, Early Ur—War Chiefs and Kings of Early Dynastic III. *Altorientalische Forschungen* 9: 43–59.
Chayanov, A. V., 1966, Peasant Farm Organization. In *The Theory of Peasant Economy*, edited by D. Thorner, B. Kerblay and R. E. F. Smith, pp. 29–269. Richard D. Irwin Publishers, Homewood.
Childe, V. G., 1925, *The Dawn of Western Civilization*. Knopf, New York.
Childe, V. G., 1950, The Birth of Civilization. *Past and Present* 2: 1–10.
Childe, V. G., 1951, *Man Makes Himself*. Mentor, New York.
Childe, V. G., 1954, *What Happened in History*. Penguin Books, Harmondsworth.
Childe, V. G., 1958a, *The Prehistory of European Society*. Penguin Books, Harmondsworth.
Childe, V. G., 1958b, Retrospect. *Antiquity* 32: 69–74.
Chochorowski, I., 1993, *Ekspansja kimmeryska na tereny europy srodkowej (Zusammenfassung: die kimmerische Expansion in das mitteleuropäsiche Gebiet)*. Uniwersytet Jagiellonski, Krakow.

REFERENCES

Clancy, F. S., 1990, A Genealogy for Freestanding Maya Monuments. In *Vision and Revision in Maya Studies*, edited by F. S. Clancy and P. D. Harrison, pp. 21–32. University of New Mexico Press, Albuquerque.

Clark, J. E., 1997, The Arts of Government in Early Mesoamerica. *Annual Review of Anthropology* 26: 211–234.

Clark, J. E., and M. Blake, 1994, The Power of Prestige: Competitive Generosity and the Origins of Rank Societies in Lowland Mesoamerica. In *Factional Competition and Political Development in the New World*, edited by E. Brumfiel and J. Fox, pp. 17–30. Cambridge University Press, Cambridge.

Clavel-Levêque, M., 1989, Puzzle Gaulois: Mémoires, Images, Textes, Histoire. *Centre de Recherches d'Histoire Ancienne*, 88. Annales Littéraires de l'Université de Besançon 396. Les Belles-Lettres, Paris.

Cleland, D. I. (ed.), 1984, *Matrix Management Systems Handbook*. Van Nostrand, New York.

Coe, W. R., and McGinn, J. J., 1963, Tikal: The North Acropolis and an Early Tomb. *Expedition* 5(2): 24–32.

Cohen, R., 1978, State Origins: A Reappraisal. In *The Early State*, edited by H. J. M. Claessen and P. Skalník, pp. 31–75. Mouton Publishers, The Hague.

Conrad, G. W., 1992, Inca Imperialism: The Great Simplification and the Accident of Empire. In *Ideology and the Evolution at the Pre-State Level: Formative Period Mesoamerica*, edited by A. A. Demarest and G. W. Conrad, pp. 159–174. School of American Research, Santa Fe, New Mexico.

Contreras Cortés, F., Cámara Serrano, J. A., Lizcano Prestel, R., Pérez Bareas, R. C., Robledo Sanz, B., and Trancho Gallo, G., 1995, Enterramientos y Diferenciación Social. I: El Registro Funerario del Yacimiento de la Edad del Bronce de Peñalosa (Baños de la Encina, Jaén). *Trabajos de Prehistoria* 52(1): 87–108.

Contreras Cortés, F., Morales Muñiz, A., Peña Chocarro, L., Robledo, B., Rodríguez Ariza, M. O., Sanz Bretón, J. L., and Trancho, G. 1995, Avance al Estudio de los Ecofactos del Poblado de Peñalosa (Baños de la Encina, Jaén): Una Aproximación a la Reconstrucción Medioambiental. In *Anuario Arqueológico de Andalucía 1992*, Volume 2: *Actividades Sistemáticas*, pp. 263–274. Consejería de Cultura de la Junta de Andalucía, Sevilla.

Cook, J., 1784, *Voyage to the Pacific Ocean in His Majesty's Ships "Resolution"and "Discovery."* Atlas, Dublin, Camerlaine.

Coombs, D., 1975, Bronze Age Weapons Hoards in Britain. *Archaeologica Atlantica* 1: 41–81.

Cordell, L. S., 1979, *Cultural Resources Overview of the Middle Rio Grande Valley, New Mexico*. U.S. Government Printing Office, Washington, DC.

Costin, C., 1986, *From Chiefdom to Empire State: Ceramic Economy Among the Pre-Hispanic Wanka of Highland Peru*. Ph.D. dissertation, Department of Anthropology, UCLA.

Costin, C., 1991, Craft Specialization: Issues in Defining, Documenting, and Explaining the Organization of Production. In *Archaeological Method and Theory*, edited by M. B. Schiffer, Volume 3, pp. 1–56. University of Arizona Press, Tucson.

Costin, C. L., and Earle, T., 1989, Status Distinctions and the Legitimation of Power as Reflected in Changing Patterns of Consumption in Late Prehispanic Peru. *American Antiquity* 54: 691–714.

Cowgill, G. L., 1983, Rulership and the Ciudadela: Political Inferences from Teotihuacan Architecture. In *Civilization in the Ancient Americas: Essays in Honor of Gordon R. Willey*, edited by R. M. Leventhal and A. L. Kolata, pp. 313–343. University of New Mexico Press, Albuquerque, and Harvard University, Cambridge.

Cowgill, G. L., 1992a, Teotihuacan Glyphs and Imagery in the Light of Some Early Colonial Texts. In *Art, Ideology, and the City of Teotihuacan*, edited by J. C. Berlo, pp. 231–246. Dumbarton Oaks, Washington, DC.

Cowgill, G. L., 1992b, Social Differentiation at Teotihuacan. In *Mesoamerican Elites: An Ar-*

chaeological Assessment, edited by D. Z. Chase and A. F. Chase, pp. 206–220. University of Oklahoma Press, Norman.

Cowgill, G. L., 1992c, Toward a Political History of Teotihuacan. In *Ideology and Pre-Columbian Civilizations*, edited by A. A. Demarest and G. W. Conrad, pp. 87–114. School of American Research Press, Santa Fe.

Cowgill, G. L., 1997, State and Society at Teotihuacan, Mexico. *Annual Review of Anthropology* 26: 129–162.

Crabtree, P. J., 1990, Zooarchaeology and Complex Societies: Some Uses of Faunal Analysis for the Study of Trade, Social Status, and Ethnicity. *Archaeological Method and Theory* 2: 155–205.

Creamer, W., 1991, *Demographic Implications of Changing Use of Space in Pueblo Settlements of Northern New Mexico During the Protohistoric, AD 1450–1680*. Paper presented at the 56th annual meeting of the Society for American Archaeology, New Orleans, LA.

Creamer, W., 1993, *The Architecture of Arroyo Hondo Pueblo*. School of American Research Press, Santa Fe.

Creamer, W., 1994, Egalitarianism, Hierarchy, and Centralization in the Pueblos. In *The Ancient Southwestern Community*, editd by W. H. Wills and R. D. Leonard, pp. 9–23. University of New Mexico Press, Albuquerque.

Creamer, W., 1996, Developing Complexity in the American Southwest: A New Model from the Rio Grande Valley. In *Emergent Social Complexity: The Evolution of Intermediate Societies*, edited by J. E. Arnold, pp. 91–106. International Models in Prehistory. University of Michigan, Ann Arbor.

Creamer, W., 2000, Regional Interactions and Regional Systems in the Protohistoric Rio Grande Valley. In *The Archaeology of Regional Interaction*, edited by M. Hegmon, pp. 99–118. University of Colorado Press, Boulder.

Creamer, W., and Haas, J., 1985, Tribe and Chiefdom in Lower Central America. *American Antiquity* 50(4): 738–754.

Creamer, W., and Haas, J., 1996, The Role of Warfare Among the Anasazi of the Pueblo III Period. In *The Prehistoric Pueblo World, AD 1150–1350*, edited by M. Adler, pp. 205–213. University of Arizona Press, Tucson.

Creamer, W., and Haas, J., 1998, Less Than Meets the Eye: Evidence for Protohistoric Chiefdoms in Northern New Mexico. In *Chiefdoms and Chieftancy in the Americas*, edited by E. M. Redmond, pp. 43–67. University of Florida Press, Gainesville.

Creamer, W., Haas, J., and Renken, L., 1994, *Testing Conventional Wisdom: Protohistoric Ceramics and Chronology in the Northern Rio Grande*. Paper presented at the 59th annual meeting of the Society for American Archaeology, Anaheim, CA.

Creamer, W., Haas, J., Burdick, D., Nelson, K. R., Renken, L., and Wenzel, A., 2000, *The Pecos Classification Doesn't Apply to Pecos: Revisiting Late Prehistoric Typology in the Northern Rio Grande*. Paper presented at the Southwest Symposium, At the Millennium: Change and Challenge in the Greater Southwest, January 14–15, 2000.

Cribb, R., 1990, *Nomads in Archaeology*. Cambridge University Press, Cambridge.

Crumley, C. L., 1979, Three Locational Models: An Epistemological Assessment for Anthropology and Archaeology. In *Advances in Archaeological Method and Theory*, edited by M. B. Schiffer, pp. 141–173.

Crumley, C. L., 1987a, A Dialectical Critique of Hierarchy. In *Power Relations_and State Formation*, edited by T. C. Patterson and C. Ward Gailey, pp. 155–168. American Anthropological Association, Washington, DC.

Crumley, C. L., 1987b, Celtic Settlement before the Conquest: The Dialectics of Landscape and Power. In *Regional Dynamics: Burgundian Landscapes in Historical Perspective*. Academic Press, San Diego.

REFERENCES

Crumley, C. L., 1993, Analyzing Historic Ecotonal Shifts. *Ecological Applications* 3(3): 377–384.
Crumley, C. L., 1994, The Ecology of Conquest: Contrasting Agropastoral and Agricultural Societies' Adaptation to Climatic Change. In *Historical Ecology*, edited by C. L. Crumley, pp. 183–201. School of American Research Press, Santa Fe.
Crumley, C. L., 1995a, Heterarchy and the Analysis of Complex Societies. In *Heterarchy and the Analysis of Complex Societies*, edited by R. M. Ehrenreich, C. L. Crumley, and J. E. Levy, Archaeological Papers of the American Anthropological Association No. 6, pp. 1–5. American Anthropological Association, Washington, DC.
Crumley, C. L., 1995b, Building an Historical Ecology of Gaulish Polities. In *Celtic Chiefdom, Celtic State*, edited by B. Gibson & B. Arnold, pp. 26–33. Cambridge University Press, Cambridge.
Crumley, C. L., 2001, From Garden to Globe: Linking Time and Space with Meaning and Memory. In *The Way the Wind Blows: Climate, History and Human Action*, edited by R. J. McIntosh, S. K. McIntosh, and Joseph A. Tainter, pp. 193–208. Columbia University Press, New York.
Crumley, C. L., and Marquardt, W. H. (eds.), 1987, *Regional Dynamics: Burgundian Landscapes in Historical Perspective*. Academic Press, San Diego.
Cuenca Payá, A., and Walker, M. J., 1986, Aspectos Paleoclimáticos del Eneolítico Alicantino. In *El Eneolítico en el País Valenciano*, pp. 43–49. Diputación de Alicante, Alicante.
Culbert, T. P., 1988, Political History and the Decipherment of Maya Glyphs. *Antiquity* 62:135–152.
Culbert, T. P. (ed.), 1991, *Classic Maya Political History: Hieroglyphic and Archaeological Evidence*. Cambridge University Press, Cambridge.
Cummins, T., 1984, Kinshape: The design of the Hawaiian feather cloaks. *Art History* 7: 1–20.
Curvers, H., and Schwartz, G., 1990, Excavations at Tell el Raqa'i: A Small Rural Site of Early Urban Northern Mesopotamia. *American Journal of Archaeology* 94(1): 3–23.
D'Altroy, T. N., 1994, *Provincial Power in the Inka Empire*. Smithsonian Institution Press, Washington, DC.
D'Altroy, T., and Earle, T., 1985, Staple finance, wealth finance, and storage in the Inca political economy. *Current Anthropology* 26: 187–206.
Davidovic, V., 1987, The Womens Ration System at Ebla. *Oriens Antiquus* 26: 299–307.
Davis, S. M., and Lawrence, P. R., 1977, *Matrix*. Addison-Wesley, Reading, MA.
De Greene, K. B., 1996, Field-Theoretic Framework for the Interpretation of the Evolution, Instability, Structural Change, and Management of Complex Systems. In *Chaos Theory in the Social Sciences: Foundations and Applications*, edited by L. D. Kiel and E. Elliott, University of Michigan Press, Ann Arbor.
Deimel, A., 1931, Sumerische Tempelwirtschaft zur Zeit Urukaginas und Seiner Vorganger. *Analecta Orientalia* 2: 71–113.
de Laguna, F., 1983, Aboriginal Tlingit Sociopolitical Organization. In *The Development of Political Organization in Native North America*, edited by E. Tucker and M. Fired, pp. 71–85. American Ethnological Society, Washington, DC.
Demarest, A. A., 1992, Ideology in Ancient Maya Cultural Evolution: The Dynamics of Galactic Polities. In *Ideology and Pre-Columbian Civilizations*, edited by A. A. Demarest and G. W. Conrad, pp. 135–157. School of American Research Press, Santa Fe, New Mexico.
DeMarrais, E., n.d., The Architecture of Xauxa Communities in Empire and Domestic Economy. In *Transformation in Household Economies of Xauxa Under the Inka*, edited by T. D'Altroy and C. Hastorf,
DeMarrais, E., Castillo, L. J., and Earle, T., 1996, Ideology, Materialization and Power Strategies. *Current Anthropology* 37: 15–31.

REFERENCES

Denton, G., and Karlen, W., 1973, Holocene Climatic Variations—Their Pattern and Possible Cause. *Quaternary Research* 3(2): 155–205.
Diakonoff, I. M., 1972, The Rural Community in the Ancient Near East. *Journal of the Economic and Social History of the Orient* 18: 121–133.
Diakonoff, I. M., 1972, Slaves, Helots, and Serfs in Early Antiquity. In *Wirtschaft und Gesellschaft im Alten Vorderasien*, edited by J. Harmatta and G. Kovoroczy, pp. 45–78. Akademiai Kiado, Budapest.
Diakonoff, I. M., 1974, *Structure of Society and State in Early Dynastic Sumer*. Undena, Malibu.
Dolce, R., 1988, Some Aspects of the Primary Economic Structures of Ebla in the Third and Second Millenniums BC: Stores and Workplaces. In *Wirtschaft und Gesellschafe von Ebla*, edited by H. Waetzoldt and H. Hauptmann, p. 35–45. Heidelberger Orientverlag, Heidelberg.
Downing, T. E., and Gibson, M. (eds.), 1974, *Irrigation's Impact on Society*. Anthropological Papers of the University of Arizona, No. 25, Tucson.
Drennan, R. D., 1991, Pre-Hispanic Chiefdom Trajectories in Mesoamerica, Central America, and Northern South America. In *Chiefdoms: Power, Economy, and Ideology*, edited by T. Earle, pp. 263–287. Cambridge University Press, Cambridge.
Drews, R., 1993, *The End of the Bronze Age. Changes in Warfare and the Catastrophe ca. 1200 BC*. Princeton University Press, Princeton.
von den Driesch, A., 1972, Osteoarchäologische Untersuchungen auf der Iberischen Halbinsel. In *Studien über frühe Tierknochenfunde von der Iberischen Halbinsel*, Volume 3. Institut für Palaeoanatomie, Domestikationsforschung und Geschichte der Tiermedizin der Universität München, Munich.
von den Driesch, A., 1973, Tierknochenfunde aus der frühbronzezeitlichen Gräberfeld von "El Barranquete," Provinz Almería. *Saügetierkundliche Mitteilungen* 21: 328–335.
Drucker, P., 1955, *Indians of the Northwest Coast*. McGraw-Hill, New York.
Dunnell, R. C., 1980, Evolutionary Theory and Archaeology. *Advances in Archaeological Method and Theory* 3: 35–99.
Dunnell, R. C., 1989, Aspects of the Application of Evolutionary Theory in Archaeology. *Archaeological Thought in America*, edited by C. C. Lamburg-Karlovsky, pp. 35–49. Cambridge University Press, Cambridge.
Durand, J. M. (ed.), 1987, *La Femme dans le Proche-Orient Antique*. Editions Recherches sur les Civilisations, Paris.
Earle, T., 1978, Economic and social organization of a complex chiefdom, the Halelea district Kaua'i,Hawaii. *Museum of Anthropology, University of Michigan, Anthropological Papers*, 63. University of Michigan, Ann Arbor.
Earle, T., 1987a, Specialization and the Production of Wealth: Hawaiian Chiefdoms and the Inka Empire. In *Specialization, Exchange, and Complex Societies*, edited by E. M. Brumfiel and T. K. Earle, pp. 64–75. Cambridge University Press, Cambridge.
Earle, T., 1987b, Chiefdoms in Archaeological and Ethnohistorical Perspective. *Annual Review of Anthropology* 16: 279–308.
Earle, T., 1990, Style and Iconography as Legitimation in Complex Chiefdoms. In *The Use of Style in Archaeology*, edited by M. Conkey and C. Hastorf, pp. 73–81. Cambridge University Press, Cambridge.
Earle, T., 1991a, The Evolution of Chiefdoms. In *Chiefdoms: Power, Economy, and Ideology*, edited by T. Earle, pp. 1–15. Cambridge University Press, Cambridge.
Earle, T., 1991b, *Chiefdoms: Power, Economy, and Ideology*. Cambridge University Press, Cambridge.
Earle, T., 1991c, Property rights and the evolution of chiefdoms. In *Chiefdoms: Power, Economy and Ideology*, edited by T. Earle, pp. 71–99. Cambridge University Press, Cambridge.

REFERENCES

Earle, T., 1994, Wealth Finance in the Inka Empire: Evidence from the Calchaqua Valley, Argentina. *American Antiquity* 59: 443–460.
Earle, T., 1997, *How Chiefs Come to Power: The Political Economy in Prehistory*. Stanford University Press, Stanford.
Earle, T., 1998, Property Rights and the Evolution of Hawaiian Chiefdoms. In *Property in Economic Contexts*, edited by R. Hunt and A. Gilman, pp. 89–118. Monographs in Economic Anthropology. University Press of America, Lanham, MD.
Earle, T. K., (ed.), 1991, *Chiefdoms: Power, Economy, and Ideology*. Cambridge University Press, Cambridge.
Earle, T. K., D'Altroy, T., Hastorf, C., Scott, C. Costin, C., Russell, G., and Sandefur, E., 1987, *Archaeological Field Research in the Upper Mantaro, Peru, 1982–1983: Investigations of Inka Expansion and Exchange*. Monograph XXVIII Institute of Archaeology, University of California, Los Angeles.
Earle, T., Bech, J., Kristiansen, K., Aperlo, P., Kelertas, K., and Steinberg, J., 1998, The political economy of Late Neolithic and Early Bronze Age Society: The Thy Archaeological Project. *Norwegian Archaeological Review* 31: 1–28.
Ehrenreich, R. M., Crumley, C. L., and Levy, J. E. (eds.), 1995, *Heterarchy and the Analysis of Complex Societies*. Archaeological Papers of the American Anthropological Association No. 6. American Anthropological Association, Washington, DC.
Eiroa, J. J., 1989, *Urbanismo protohistórico de Murcia y el sureste*. Universidad de Murcia, Murcia.
Eliade, M., 1970, *Shamanism: Archaic Techniques of Ecstasy*. Translated from French by Willard R. Trask. Princeton University Press, Princeton.
Ellen, R., 1982, *Environment, Subsistence, System: The Ecology of Small Scale Social Formations*. Cambridge University Press, Cambridge.
Ellis, F. H., 1951, Patterns of Aggression and the War Cult in Southwestern Pueblos. *Southwestern Journal of Anthropology* 7: 177–201.
Ellis, R., and Voigt, M., 1982, Excavations at Gritille, Turkey. *American Journal of Archaeology* 86: 319–332.
Ember, M., and Ember, C., 1994, Cross-Cultural Studies of War and Peace: Recent Achievements and Future Possibilities. In *Studying War: Anthropological Perspectives*, edited by S. P. Reyna and R. E. Downs, pp. 185–208. Gordon and Breach, Amsterdam.
Engel, F., 1963, A Preceramic Settlement on the Central Coast of Peru: Asia Unit 1. *American Philosophical Society Transaction*, New Series 53(3): 1–139.
Engels, F., 1972, *The Origin of the Family, Private Property and the State*. International Publishers, New York.
Enright, M. J., 1996, *Lady with a Mead Cup: Ritual, Prophecy and Lordship in the European Warband from La Tene to the Viking Age*. Four Courts Press, Portland.
Falkenstein, A., 1974, *The Sumerian Temple City*. Maria deJ. Ellis, translator. Volume 1/1. Undena Publications, Malibu.
Fairservis, W. A. Jr., 1961, *The Harappan Civilization: New Evidence and More Theory*. Novitates No. 2055. American Museum of Natural History, New York.
Fairservis, W. A. Jr., 1967, *The Origin, Character, and Decline of an Early Civilization*. Novitates No. 2302. American Museum of Natural History, New York.
Feeley-Harnik, G., 1985, Issues in Divine Kingship. *Annual Review of Anthropology* 14: 273–313.
Feinman, G. M., 1991, Demography, Surplus, and Inequality: Early Political Formations in Highland Mesoamerica. In *Chiefdoms: Power, Economy, and Ideology*, edited by T. Earle, pp. 229–262. Cambridge University Press, Cambridge.
Feinman, G. M., 1995, The Emergence of Inequality: A Focus on Strategies and Processes. In

Foundations of Social Inequality, edited by T. D. Price and G. M. Feinman, pp. 255–279. Plenum Press, New York.

Feinman, G. M., 2000, Corporate/Network: A New Perspective on Leadership in the American Southwest. In *Hierarchies in Action: Cui Bono?* edited by M. Diehl, pp. 152–180. Occasional Paper o. 27. Center for Archaeological Investigations, Southern Illinois University, Carbondale.

Feinman, G. M., and Neitzel, J., 1984, Too Many Types: An Overview of Sedentary Prestate Societies in the Americas. In *Advances in Method and Theory,* Volume 7, edited by M. S. Schiffer, pp. 39–102. Academic Press, New York.

Feldman, R. A., 1987, Architectural Evidence for the Development of Nonegalitarian Social Systems in Coastal Peru. In *The Development and Origins of the Andean State*, edited by J. Haas, S. Pozorski, and T. Pozorski, pp. 9–14. Cambridge University Press, Cambridge.

Ferguson, B. R., and Whitehead, N. (eds.), 1992, *War in the Tribal Zone: Expanding States and Indigenous Warfare*. School of American Research Press, Santa Fe.

Fernández-Miranda, M., Fernàndez-Posse, D. M., Gilman, A., and Martín, C., 1993, El Sustrato Neolítico en la Cuenca de Vera (Almería). *Trabajos de Prehistoria* 50: 57–85.

Fernández-Posse, D. M., Gilman, A., & Martín, C., 1996, Consideraciones Cronológicas Sobre la Edad del Bronce en La Mancha. In *Homenaje al profesor Manuel Fernández-Miranda*, edited by M. Angeles Querol and T. Chapa, Vol. 2, pp. 111–137. Servicio de Publicaciones, Universidad Complutense, Madrid.

Flannery, K. V., 1972, The Cultural Evolution of Civilizations. *Annual Review of Ecology and Systematics* 3: 399–426.

Flannery, K. V., 1998, The Ground Plans of Archaic States. In *Archaic States*, edited by G. M. Feinman and J. Marcus, pp. 15–57. School of American Research Press, Santa Fe.

Fleener, M. J., and Pourdavood, R. G., 1997, *Autopoesis and Change: Ontological Considerations and Implications of School Reform*. In press.

Fleming, A., 1973, Tombs for the living. *Man* 8: 177–192.

Fleming, A., 1982, Social boundaries and land boundaries. In *Ranking, Resources and Exchange*, edited by C. Renfrew and S. Shennan, pp. 52–55. Cambridge University Press, Cambridge.

Fleming, A., 1989, The genesis of coaxial field systems. In *What's New? A Closer Look at the Process of Innovation*, edited by S. E. van der Leeuw and R. Torrence, pp. 63–81. Unwin Hyman, London.

Fletcher, R., 1995, *The Limits of Settlement Growth: A Theoretical Outline*. Cambridge University Press, Cambridge.

Florschütz, F., Menéndez Amaor, J., & Wijmstra, T. A., 1971, Palynology of a Thick Quaternary Succession in Southern Spain. *Palaeogeography, Palaeoclimatology, Palaeoecology* 10: 223–264.

Folan, W. J., Marcus, J., and Miller, W. F., 1995a, Verification of a Maya Settlement Model through Remote Sensing. *Cambridge Archaeological Journal* 5(2): 277–301.

Folan, W. J., Marcus, J., Pinceman, S., delR. Dominguez Carrasco, M., Fletcher, L., and Morales López, A, 1995b, Calakmul: New Data from an Ancient Maya Capital in Campeche, Mexico. *Latin American Antiquity* 6: 310–334.

Ford, R. I., 1972, Barter, Gift, or Violence: An analysis of Tewa Intertribal Exchange. In *Social Exchange and Interaction*, edited by E. N. Wilmsen, pp. 21–45. Museum of Anthropology, University of Michigan Anthropological Papers 46, Ann Arbor.

Foster, B., 1977, Commercial Activity in Sargonic Mesopotamia. *Iraq* 39: 31–43.

Foster, B., 1981, A New Look at the Sumerian Temple State. *Journal of the Economic and Social History of the Orient* 24(3): 225–241.

Fox, R. G. (ed.), 1991, *Recapturing Anthropology: Working in the Present*. School of American Research Press Santa Fe, 1991.

REFERENCES

Freidel, D. A., 1981, The Political Economics of Residential Dispersion Among the Lowland Maya. In *Lowland Maya Settlement Patterns*, edited by W. Ashmore, pp. 371–382. University of New Mexico Press, Albuquerque.

Freidel, D. A., 1992, The Trees of Life: Ahau as Artifact in Classic Lowland Maya Civilization. In *Ideology and Pre-Columbian Civilizations*, edited by A. A. Demarest and G. W. Conrad, pp. 115–133. School of American Research Press, Santa Fe, New Mexico.

Freidel, D. A., and Schele, L., 1988, Kingship in the Late Preclassic Maya Lowlands: The Instruments and Places of Ritual Power. *American Anthropologist* 90: 547–567.

Freidel, D., Schele, L., and Parker, J. 1993, *Maya Cosmos: Three Thousand Years on the Shaman's Path*. William Morrow and Co., New York.

Freitag, H., 1971, Die natürliche Vegetation des südostspanischen Trockengebietes. *Botanische Jahrbücher* 91: 147–308.

Fried, M. H., 1960, On the Evolution of Social Stratification and the State. In *Culture in History*, edited by S. Diamond, pp. 713–731. Columbia University Press, New York.

Fried, M. H., 1967, *The Evolution of Political Society: An Essay in Political Anthropology*. Random House, New York.

Fried, M. H., 1968, State: The Institution. In *International Encyclopedia of the Social Sciences* 15:143–150. Macmillan and the Free Press, New York.

Friedman, J., 1982, Catastrophe and Continuity in Social Evolution. In *Theory and Explanation in Archaeology: The Southampton Conference*, edited by C. Renfrew, M. J. Rowlands, and B. A. Segraves, pp. 175–196. Academic Press, New York.

Friedman, J., and Rowlands, M. J., 1977, Notes towards an epigenetic model of the evolution of "civilization." In *The Evolution of Social Systems*, edited by J. Friedman and M. Rowlands, pp. 201–276. Duckworth, London.

Furst, P. T., 1974, The Roots and Continuities of Shamanism. *Artscanada* 184–187: 33–60.

Futrell, M., 1998, Social Boundaries and Interaction: Ceramic Zones in the Northern Rio Grande Pueblo IV Period. In *Migration and Reorganization: The Pueblo IV Period in the American Southwest*. Arizona State University Anthropological Research Papers 51: 285–292.

Gailey, C., and Patterson, T., 1987, Power Relations and State Formation. In *Power Relations and State Formation*, edited by T. Patterson and C. Gailey, pp. 1–26. American Anthropological Association, Washington, DC.

Garelli, P. (ed.), 1974, *Le Palais et la Royaute*. P.Guethner, Paris.

Geiger, F., 1972, Die Bewässerungswirtschaft Südostspaniens im trockensten Abschnitt des Mediterrane Europas. *Geographische Rundschau* 24: 408–419.

Gelb, I., 1965, The Mesopotamian Ration System. *Journal of Near Eastern Studies* 24(3): 230–243.

Gelb, I., 1969, On the Alleged Temple and State Economies in Ancient Mesopotamia. In *Studi in Onore di Edouardo Volterra*, Volume 6, pp. 137–154, Giuffre Editore, Milan.

Gelb, I., 1973, Prisoners of War in Early Mesopotamia. *Journal of Near Eastern Studies* 32(1): 70–98

Gelb, I., Steinkeller, P., and Whiting, R., 1991, *Earliest Land Tenure Systems in the Ancient Near East: Ancient Kudurrus*. University of Chicago Oriental Institute, Chicago.

Gent, H., 1983, Centralised storage in later prehistoric Britain. *Proceedings of the Prehistoric Society* 49: 243–467.

Gerloff, S., 1995, Bronzezeitliche Goldblechkronen aus Westeuropa. Betrachtungen zur Funktion der Goldblechkegel vom Typ Schifferstadt und der atlantischen"Goldschalen" der Form Devil's Bit und Atroxi. In *Festschoift fur Hermann Muller Karpe* zum 70, edited by A. Jockenhovel Gebertsstag pp. 153–194). Rudolf Hubelt, Bonn.

Gibson, M., 1974, Violation of Fallow and Engineered Disaster in Mesopotamian Civilization. In *Irrigation's Impact on Society*, edited by T. Downing and M. Gibson, pp. 7–20. University of Arizona, Tucson.

Giddens, A., 1984, *The Constitution of Society: Outline of the Theory of Structuration*. University of California Press, Berkeley.
Gillespie, S. D., 1993, Power, Pathways, and Appropriations in Mesoamerican Art. In *Imagery and Creativity: Ethnoaesthetics and Art Worlds in the Americas*, edited by D. S. Whitten and N. E. Whitten, Jr., pp. 67–107. University of Arizona Press, Tucson.
Gilman, A., 1976, Bronze Age Dynamics in Southeast Spain. *Dialectical Anthropology* 1: 307–319.
Gilman, A., 1981, The Development of Social Stratification in Bronze Age Europe. *Current Anthropology* 22: 1–23.
Gilman, A., 1987a, El Análisis de Clase en la Prehistoria del Sureste. *Trabajos de Prehistoria* 44: pp. 27–34.
Gilman, A., 1987b, Regadío y conflicto en sociedades acéfalas. *Boletín del Seminario de Arte y Arqueología, Universidad de Valladolid*, 53: pp. 59–72.
Gilman, A., 1987c, Unequal Exchange in Copper Age Iberia. In *Specialization, Exchange, and Complex Society*, edited by E. M. Brumfiel and T. K. Earle, pp. 22–29. Cambridge University Press, Cambridge.
Gilman, A., 1991, Trajectories Towards Social Complexity in the Later Prehistory of the Mediterranean. In *Chiefdoms: Power, Economy, and Ideology*, edited by T. Earle, pp. 146–169. Cambridge University Press, Cambridge.
Gilman, A., 1995a, Prehistoric European Chiefdoms: Rethinking "Germanic" Societies. In *Foundations of Social Inequality*, edited by T. D. Price & G. M. Feinman, pp. 235–251. Plenum Press, New York.
Gilman, A., 1995b, Recent Trends in the Archaeology of Spain. In *The Origins of Complex Societies in Late Prehistoric Spain*, edited by K. T. Lillios, pp. 1–6. International Monographs in Prehistory, Ann Arbor.
Gilman, A., & Thornes, J. B., 1985, *Land-Use and Prehistory in South-East Spain*. George Allen & Unwin, London.
Gledhill, J., Bender, B., and Trolle Larsen, M. (eds.), 1988, *State and Society: The Emergence and Development of Social Hierarchy and Political Centralization*. Unwin Hyman, London.
Gleick, J., 1988, *Chaos: Making a New Science*. Viking, New York.
Glob, P. V., 1944, Studier over den jyske Enkelt gravskulter. *Aarboger for nordisk Oldkyndighed og Historie*. Det kgl. Nordiske Oldskriftsselskab, Kobenhaven.
González Prats, A., Ruiz Segura, E., Gil Fuensanta, J., and Seva Román, R., 1998, Cerámica Roja Monocroma Anatólica en el Poblado Calcolítico de Les Moreres (Crevillente, Alicante, España). *Lucentum* 11–13 (1992-1994): 7–38.
Goodwin, B., 1994, *How the Leopard Changed its Spots: The Evolution of Complexity*. Weidenfeld & Nicholson, London.
Green, M. W., 1981, The Construction and Implementation of the Cuneiform Writing System. *Visible Language* 15: 345–372.
Green, S. W., and Perlman, S. M., (eds.), 1985, *The Archaeology of Frontiers and Boundaries*. Academic Press, New York.
Gregg, S. A. (ed.), 1991, *Between Bands and States*. Center for Archaeological Investigations, Southern Illinois University at Carbondale, Occasional Papers No. 9.
Grove, D., and Gillespie, S. D., 1992, Ideology and the Evolution at the Pre-State Level: Formative Period Mesoamerica. In *Ideology and Pre-Columbian Civilizations*, edited by A. A. Demarest and G. Conrad, pp. 15–36. School of American Research Press, Santa Fe.
Gruzinski, S., 1989, *Man-Gods in the Mexican Highlands: Indian Power and Colonial Society, 1520–1800*. Translated from French by E. Corrigan. Stanford University Press, Stanford.
Guilaine, J., 1994, *La Mer Partagée: La Méditerranée avant l'Écriture, 7000–2000 avant Jésus-Christ*. Hachette, Paris.

REFERENCES

Gumerman, G. J. (ed.), 1988, *The Anasazi in a Changing Environment*. Cambridge University Press, Cambridge and New York.

Gumerman, G. J., and Kohler, T. A., 1994, Archaeology and the Sciences of Complexity. Paper presented at the symposium *Prehistoric Cultures as Complex Adaptive Systems*. M. Diehl and G. Gumerman, organizers. Annual Meeting, Society for American Archaeology, April, Anaheim.

Gunn, J. D., 1994, Global Climate and Regional Biocultural Diversity. In *Historical Ecology: Cultural Knowledge and Changing Landscapes*, edited by C. L. Crumley, pp. 67–97. School of American Research, Santa Fe.

Gunn, J. D., and Grzymala-Busse, J. W., 1994, Global Temperature Stability by Rule Induction: An Interdisciplinary Bridge. *Human Ecology* 22(1): 59–81.

Haas, J., 1981, Class Conflict and the State in the New World. In *The Transition to Statehood in the New World*, edited by G. D. Jones and R. R. Kautz, pp. 80–102. Cambridge University Press, Cambridge.

Haas, J., 1982, *The Evolution of the Prehistoric State*. Columbia University Press, New York.

Haas, J., 1987, The Exercise of Power in Early Andean State Development. In *The Development and Origins of the Andean State*, edited by J. Haas, S. Pozorski, and T. Pozorski, pp. 31–35. Cambridge University Press, Cambridge.

Haas, J., 1989, The Evolution of the Kayenta Regional System. In *The Sociopolitical Structure of Prehistoric Southwestern Societies*, edited by S. Upham and K. Lightfoot. Westview Press, Boulder, CO.

Haas, J., 1990, Warfare and the Evolution of Tribal Polities in the Prehistoric Southwest. In *The Anthropology of War*, edited by J. Haas, pp. 171–189. Cambridge University Press, Cambridge.

Haas, J., 1998b, Warfare and the Evolution of Culture. In *Working Papers, 98-11-088*. Santa Fe Institute, Santa Fe.

Haas, J., and Creamer, W., 1993, Stress and Warfare Among the Kayenta Anasazi of the Thirteenth Century A.D. *Fieldiana Anthropology*, New Series, No. 21. Field Museum of Natural History, Chicago.

Haas, J., and Creamer, W., 1996, The Role of Warfare Among the Anasazi of the Pueblo III Period. *In The Prehistoric Pueblo World, AD 1150–1350*. Edited by Michael Adler, pp. 205–213. University of Arizona Press.

Haas, J., and Creamer, W., 1998, *Pueblo Political Organization in 1500: Tinkering With Diversity*. Paper presented at the 63rd annual meeting of the Society for American Archaeology, Seattle.

Habicht-Mauche, J. A., 1988, *Town and Province: Regional Integration among the Classic Period Rio Grande Pueblos*. Paper presented at the 53rd Annual Meeting of the Society of American Archaeology, Phoenix.

Habicht-Mauche, J. A., 1995, Changing Patterns of Pottery Manufacture and Trade in the Northern Rio Grande Region. In *Ceramic Production in the American Southwest*, edited by B. J. Mills and P. L. Crown, pp. 167–199. University of Arizona Press, Tucson.

Habicht-Mauche, J. A., Geselowitz, M., and Hoopes, J., 1987, *Where's the Chief? The Archaeology of Complex Tribes*. Paper Presented at the 52nd Annual Meeting of the Society for American Archaeology. Toronto, Ontario.

Hägg, R., 1984, Degrees and character of the Minoan influence on the mainland. In *The Minoan Thalassocracy: Myth and Reality*, edited by R. Hägg and N. Marinatos, Skrifter utgivna av svenska institutet i Athen, 4, XXXII. Stockholm.

Hagstrum, M., 1985, Measuring Prehistoric Ceramic Craft Specialization: A Test Case in the American Southwest. *Journal of Field Archaeology* 12: 65–75.

Hagstrum, M., 1985, 1986, The Technology of Ceramic Production of Wanka and Inka Wares from the Yanamarca Valley, Peru. *Ceramic Notes* 3: 1–29.

Haken, H., 1983, *Advanced Synergetics: Instability Hierarchies of Self-Organizing Systems and Devices*. Springer-Verlag, Berlin.

Hammond, G. P., and Rey, A., 1940, *Narratives of the Coronado Expedition*. University of New Mexico Press, Albuquerque.

Hammond, G. P., and Rey, A., 1966, *The Rediscovery of New Mexico*. University of New Mexico Press, Albuquerque.

Hammond, N. (ed.), 1991, *Cuello: An Early Maya Community in Belize*. Cambridge University Press, Cambridge.

Hanks, W., 1990, *Referential Practice: Language and Lived Space Among the Maya*. University of Chicago Press, Chicago.

Hansen, R. D., 1991, The Road to Nakbe. *Natural History* 5/91(5): 8–14.

Hansen, R. D., 1998, Continuity and Disjunction: The Preclassic Antecedents of Classic Architecture. In *Function and Meaning in Classic Maya Architecture*, edited by S. D. Houston, pp. 49–122. Dumbarton Oaks Research Library and Collection, Washington, D C.

Harris, M., 1968, *The Rise of Anthropological Theory*. Crowell, New York.

Harrison, S., 1987, Magical and Material Polities in Melanesia. *Man* 24: 1–20.

Harvey, D. L., Reed, M., 1996, Social Science as the Study of Complex Systems. In edited by L. D. Kiel and E. Elliott, *Chaos Theory in the Social Sciences: Foundations and Applications*. University of Michigan Press, Ann Arbor.

Hassig, R., 1988, *Aztec Warfare: Imperial Expansion and Political Control*. University of Oklahoma Press, Norman.

Hassig, R., 1992, *War and Society in Ancient Mesoamerica*. University of California Press, Berkeley.

Hastorf, C., 1993, *Agriculture and the Onset of Political Inequality Before the Inka*. Cambridge University Press, Cambridge

Hastorf, C., and Earle, T., 1985, Intensive agriculture and the geography of political change in the Upper Mantaro region of Central Peru. In *Prehistoric Intensive Agriculture in the Tropics*, edited by I. S. Farrington. BAR International Series 232.

Haviland, W. A., 1992, From Double Bird to Ah Cacao: Dynastic Troubles and the Cycle of Katuns at Tikal, Guatemala. In *New Theories on the Ancient Maya*, edited by E. C. Danien and R. J. Sharer, pp. 71–80. University Museum Monograph 77, Philadelphia.

Hawkes, C. C., 1954, Archaeological Theory and Method: Some Suggestions from the Old World. *American Anthropologist* 56" 155–168.

Hayden, B., 1990, Nimrods, Piscators, Pluckers, and Planters: The Emergence of Food Production. *Journal of Anthropological Archaeology* 9: 31–69.

Hayden, B., 1995, Pathways to Power: Principles for Creating Socioeconomic Inequalities. In *Foundations of Social Inequality*, edited by T. D. Price and G. M. Feinman, pp. 15–86. Plenum Press, New York.

Hedeager, L., 1997, *Skygger af en anden virkelighed. Oldnordiske myter*. Samlerens Universitet, Kobenhaven.

Helms, M., 1979, *Ancient Panama: Chiefs in Search of Power*. University of Texas Press, Austin.

Helms, M., 1988, *Ulysses'Sail. An Ethnographic Odyssey of Power, Knowledge, and Geographical Distance*. Princeton University Press, Princeton.

Helms, M., 1993, *Craft and the Kingly Ideal. Art, Trade and Power*. University of Texas Press, Austin.

Helms, M. W., 1999, Why Maya Lords Sat on Jaguar Thrones. In *Material Symbols: Culture and Economy in Prehistory*, edited by J. E. Robb. Occasional Paper No. 26, Center for Archaeological Investigations, Southern Illinois University, Carbondale.

REFERENCES

Hernando Gonzalo, A., 1987, ¿Evolución Cultural Diferencial del Calcolítico Entre las Zonas Áridas y Húmedas del Sureste Español? *Trabajos de Prehistoria* 44: pp. 171–200.

Herrmann, J., 1982, Militärische Demokratie und die übergangsperiode zur Klassengesellschaft. *Etnographisch-Archäologische Zeitschrift*, 23:11–31.

Hill, R. E., and White, B. J., 1979, *Matrix Organization & Project Management*. Michigan Business Papers Number 64. Division of Research, Graduate School of Business Administration, University of Michigan, Ann Arbor.

Hodder, I., 1984, Burials, Houses, Women and Men in the European Neolithic. In *Ideology, Power and Prehistory*, edited by D. Miller, and C. Tilley. Cambridge University Press, Cambridge.

Hole, F., 1981, The Prehistory of Herding: Some Suggestions from Ethnography. In *L'Archeologie de l'Iraq du Debut de l'Epoque Neolithique a 333 Avant Notre Ere. Perspectives et Limites de l'Interpretation Anthropologique des Documents*, edited by M. T. Barrelet, pp. 119–130. Colloque Internationale No. 580, Paris. CNRS.

Holling, C. S., 1973, Resilience and Stability of Ecological Systems. *Annual Review of Ecology and Systematics* 4: 1–23.

Houston, S., and Stuart, D., 1996, Of Gods, Glyphs, and Kings: Divinity and Rulership among the Classic Maya. *Antiquity* 70: 289–312.

Houston, S., and Stuart, D., 2000, Peopling the Maya Coast. In *Peopling the Classic Maya Court*, edited by T. Inonata and S. D. Houston, Westview Press, Boulder.

Houston, S., Robertson, J., and Stuart, D., 2000, *The Language of Classic Maya Inscriptions*. *Current Anthropology 41* (3): 321–356.

Howell, T. L., 1994, The Decision-Making Structure of Protohistoric Zuni Society. In *Exploring Social, Political, and Economic Organization in the Zuni Region*, edited by T. L. Howell and T. Stone, pp. 61–90. Anthropological Research Papers No. 46, Arizona State University, Tempe.

Hunter-Anderson, R., 1979, Explaining residential aggregation in the northern Rio Grande: A competition reduction model. In *Archaeological Investigations in Cochiti Reservoir* 4, edited by Jan V. Biella and Richard C. Chapman, pp. 169–175. Office of Contract Archaeology, University of New Mexico, Albuquerque.

Jacob, F., 1977, Evolution and Tinkering. *Science* 196 (4295): 1161–1166.

Jacobsen, T., 1943, Primitive Democracy in Ancient Mesopotamia. *Journal of Near Eastern Studies* 2: 159–172.

Jacobsen, T., 1957, Early Political development in Mesopotamia. *Zeitschrift für Assyriologie* 52: 91–140.

Jacobsen, T., 1982, *Salinity and Irrigation Agriculture in Antiquity*. Bibliotheca Mesopotamica 14, Malibu.

Janger, A. R., 1963, Anatomy of the Project Organization. *The Conference Business Management Record*, November 1963.

Janger, A. R., 1979, *Matrix Organization of Complex Businesses*. The Conference Board, New York.

Jantsch, E., 1982, From Self-Reference to Self-Transcendence: The Evolution of Self-Organization Dynamics. In *Self-Organization and Dissipative Structures: Applications in the Physical and Social Sciences,* edited by W. C. Schieve and P. M. Allen. University of Texas Press, Austin.

Jiménez Brebeil, S., and García Sánchez, M., 1990, Estudio de los Restos Humanos de la Edad del Bronce del Cerro de la Encina (Monachil, Granada). *Cuadernos de Prehistoria de la Universidad de Granada* 14–15: 157–180.

Jockenhövel, A., 1990, Bronzezeitliche Burgenbau in Mitteleuropa. Untersuchungen zur Struktur frühmetallzeitlicher Gesellschaften. In *Orientalisch-ägäische Einflüsse in der europäischen*

REFERENCES

Bronzezeit, Römisch-Germanisches Zentralmuseum, Monographien Band 15. Rudolf Habelt, Bonn.

Johnson, A., and Earle, T., 1987, *The Evolution of Human Societies*. Cambridge University Press, Cambridge.

Johnson, G. A., 1973, Local Exchange and Early State Development in Southwestern Iran. *Anthropological Papers*, Volume 51. Museum of Anthropology, University of Michigan. Ann Arbor.

Johnson, G. A., 1978, Information Sources and the Development of Decision-Making Organizations. In *Social Archaeology: Beyond Subsistence and Dating*, edited by C. E. A. Redman, pp. 87–112. Academic Press, New York.

Johnson, G. A., 1980, Spatial Organization of Early Uruk Settlement Systems. In *L'Archéologie de l'Iraq du Debut de l'Epoque Néolithique a 333 Avant Notre Ere. Perspectives et Limites de l'Interpretation Anthropologique des Documents*, edited by M. T. Barrelet, pp. 233–263. Colloques Internationaux du Centre Nationale du recherche Scientifique 580. CNRS, Paris.

Johnson, G. A., 1982, Organizational Structure and Scalar Stress. In *Theory and Explanation in Archaeology*, edited by C. Renfrew, M. J. Rowlands and B. J. Seagraves, pp. 389–422. Academic Press, New York.

Johnson, G. A., 1987, The Changing Organization of Uruk Administration on the Susiana Plain. In *The Archaeology of Western Iran*, edited by F. Hole, pp. 107–139. Smithsonian Institution Press, Washington.

Johnson, G. A., 1989, Dynamics of Southwestern Prehistory: Far Outside—Looking in. In *Dynamics of Southwestern Prehistory*, edited by L. S. Cordell and G. J. Gumerman, pp. 371–389. Smithsonian Institution Press, Washington, DC.

Jones, C., 1991, Cycles of Growth at Tikal. In *Classic Maya Political History: Hieroglyphic and Archaeological Evidence*, edited by T. P. Culbert, pp. 102–127. Cambridge University Press, Cambridge.

Jones, C., and Satterthwaite, L., 1982, *The Monuments and Inscriptions of Tikal: The Carved Monuments*. Tikal Report No. 33, Part A, University Museum Monograph 44. The University Museum, University of Pennsylvania, Philadelphia.

Jones, S., 1997, *The Archaeology of Ethnicity: Constructing Identities in the Past and Present*. Routledge, London.

Joyce, A., and Winter, M., 1996, Ideology, Power, and Urban Society in Pre-Hispanic Oaxaca. *Current Anthropology* 37(1): 33–47.

Kaliff, A., 1998, Grave Structures and Altars: Archaeological Traces of Bronze Age Eschatological Conceptions. *European Journal of Archaeology*, 1 (2): 177–198.

Kauffman, S. A., 1993, *The Origins of Order: Self-Organization and Selection in Evolution*. Oxford University Press, New York.

Kauffman, S. A., 1995, *At Home in the Universe*. Oxford University Press, New York.

Kaul, F., 1998, Ships on Bronzes. A Study in Bronze Age Religion and Iconography. *Publications from the National Museum. Studies in Archaeology and History*, Vol. 3. Copenhagen.

Keeley, L. H., 1996, *War Before Civilization: The Myth of the Peaceful Savage*. Oxford University Press, New York.

Kenoyer, J. M., 1991, The Indus Valley Tradition of Pakistan and Western India. *Journal of World Prehistory* 5: 331–385.

Kenoyer, J. M., 1998, *Ancient Cities of the Indus Valley Civilization*. American Institute of Pakistan Studies, Oxford.

Kerzner, H., and Cleland, D. I., 1985, *Project/Matrix Management Policy and Strategy: Cases and Situations*. Van Nostrand, New York.

Kiel, L. D., and Elliott, E. (eds.), 1996, *Chaos Theory in the Social Sciences: Foundations and Applications*. University of Michigan Press, Ann Arbor.

REFERENCES

Kilian, K., 1988, The Emergence of Wanax Ideology in the Mycenaean Palace. *Oxford Journal of Archaeology*, (3): 291–303.

King, C., 1990, The Evolution of Chumash Society: A Comparative Study of Artifacts Used. In *Social Systems Maintenance in the Santa Barbara Channel Region before AD 1804*, pp. Garland, New York.

Klengel, H., 1996, Kultgeschehen und Symbolgut im Text zeugnis der Hethiter. In *Archaologische Forshungen zum Kultgeschehen in der jungeren Bronzezedit und fruhen Eisenreit Alteuropas*, pp. Universitat Regensburg, Regensburg.

Knight, K. (ed.), 1977, *Matrix Management*. PBI, New York.

Kohler, T. A., and Van West, C. R., 1996, The Calculus of Self- Interest in the Development of Cooperation: Sociopolitical Development and Risk among the Northern Anasazi. In *Evolving Complexity and Environment: Risk in the Prehistoric Southwest*, edited by J. A. Tainter and B. Bagley Tainter, pp. 169–196. Addison Wesley, Reading.

Kolata, A. L., 1984, The Tree, The King and the Cosmos: Aspects of Tree Symbolism in Ancient Mesoamerica. *Field Museum of Natural History Bulletin* 55(3): 10–19.

Kolb, C. C., 1987, *Marine Shell Trade and Classic Teotihuacan, Mexico*. BAR International Series 364, Oxford.

Kolb, M., 1991, *Social Power, Chiefly Authority, and Ceremonial Architecture in an Island Polity, Maui, Hawaii*. Ph.D. dissertation, Department of Anthropology, University of California at Los Angeles.

Kolb, M., 1994, Monumentality and the Rise of Religious Authority in Precontact Hawai'i Social Evolution. *Current Anthropology* 34: 1–38.

Kolb, M., 1997, Labor Mobilization, Ethnohistory, and the Archaeology of Community in Hawai'i. *Journal of Archaeological Method and Theory* 4: 265–285.

Kolb, M. J., and Snead, J. E. 1997, "It's a Small World After All: Comparative Analyses of community Organization in Archaeology." *American Antiquity* 62(1): 609–628.

Kowalewski, S. A., 1998, *Cyclical Transformations in North American Prehistory*. Paper presented at the Concepts of Humans and Behavioral Patterns in the Cultures of the East and the West: Interdisciplinary Approach Workshop. Russian State University for Humanities, Moscow.

Kramer, R. J., 1994, *Organizing for Global Competitiveness: The Matrix Design*. Report Number 1088-94-RR. The Conference Board, New York.

Kristiansen, K., 1982, The Formation of Tribal Systems in Later European Prehistory, 4000–500 BC. In *Theory and Explanation in Archaeology: The Southampton Conference*, edited by C. Renfrew, M. Rowlands and B. Seagrave, pp. 241–80. Academic Press, New York.

Kristiansen, K., 1984, Ideology and Material Culture: An archaeological Perspective. In *Marxist Perspectives in Archaeology*, edited by M. Spriggs, pp. 72–100. Cambridge University Press, Cambridge.

Kristiansen, K., 1987, From Stone to Bronze. The Evolution of Social Complexity in Northern Europe 2300–1200 BC. In *Specialization, Exchange, and Complex Societies*, edited by E. M. Brumfield and T. Earle, pp. 30–51. Cambridge University Press: Cambridge.

Kristiansen, K., 1991, Chiefdoms, States, and Systems of Social Evolution. In *Chiefdoms: Power, Economy, and Ideology*, edited by T. Earle, pp. 16–43. Cambridge University Press, Cambridge.

Kristiansen, K., 1998a, *Europe Before History*. New Studies in Archaeology. Cambridge University Press, Cambridge.

Kristiansen, K., 1998b, A Theoretical Strategy for the Interpretation of Exchange and Interaction in a Bronze Age Context. In *L'Atelier du bronzier en Europe du XXe au VIIIe siècle avant notre ère*, edited by C. Mordant, M. Pernot and V. Rychner. Actes du Colloque International "Bronze 96" Neuchatel et Dijon, 1996. Tome III (Session de Dijon) Produc-

tion, circulation et consommation du bronze. Paris.
Kristiansen, K., 1998c, The Construction of a Bronze Age Landscape. Cosmology, Economy and Social Organisation in Thy, Northwest Jutland. In *Mensch und Umwelt in der Bronzezeit Europas,* B. Hänsel (ed.). Oetkers-Voges Verlag, Kiel.
Kristiansen, K., 1999, The Emergence of Warrior Aristocracies in Later European Prehistory and their Long-Term History. In *Ancient Warfare. Archaeological Perpectives,* edited by J. Carman & A. Harding. Sutton Publishing, United Kingdom.
Kristiansen, K., and Larsson, T., *The Great Journey: Symbolic Transmission and Social Transformation in Bronze Age Europe.* (in press).
Kristiansen, K., and Rowlands, M., 1998, *Social Transformations in Archaeology. Global and Local Perspectives.* Routledge, London and New York.
Kunter, M., 1990, Menschliche Skelettreste aus Siedlungen der El Argar-Kultur. *Madrider Beiträge,* Volume 8. Philipp von Zabern, Mainz am Rhein.
Kupper, J. R., 1957, *Les Nomades en Mésopotamie au Temps des Rois de Mari.* Société d'Édition "Les Belles Lettres," Paris.
Lambert, M., 1954, *Paa-ko, Archaeological Chronicle of an Indian Village in North Central New Mexico.* Monographs of the School of American Research 19. Santa Fe.
Lambert, P. M., and Walker, P. L., 1991, Physical Anthropology and Social Complexity. *Antiquity* 65(249): 963–974.
Langton, C. G. (ed.), 1992, *Artificial Live II: Proceedings of the Workshop on Artificial Life.* Santa Fe, NM. Addison-Wesley, Redwood City.
Laporte, J. P., and Fialko, C., 1990, New Perspectives on Old Problems: Dynastic References for the Early Classic at Tikal. In *Vision and Revision in Maya Studies,* edited by F. S. Clancy and P. D. Harrison, pp. 33–66. University of New Mexico Press, Albuquerque.
Larsson, T., 1997, Materiell kultur och religiöse symboler. Mesopotamien, Anatolien och Skandinavien under det andra förkristna årtusendet. *Arkeologiska studier vid Umeå Universitet* 4.
Larsen, M. T., 1989, Introduction: Literacy and Social Complexity. In *State and Society: The Emergence and Development of Social Hierarchy and Political Centralization,* edited by J. Gledhill, B. Bender, and M. T. Larsen, pp. 173–191. Unwin Hyman, London.
Lauk, H. D., 1976, Tierknochenfunde aus bronzezeitlichen Siedlungen bei Monachil und Purullena (Provinz Granada). In *Studien über frühe Tierknochenfunde von der Iberischen Halbinsel,* Volume 6. Institut für Palaeoanatomie, Domestikationsforschung und Geschichte der Tiermedizin der Universität München, Munich.
Layton, R. A., Foley, R., and Williams, E., 1991, The Transition between Hunting and Gathering and the Specialized Husbandry of Resources: A Socio-ecological Approach. *Current Anthropology* 32(3): 255–274.
Leach, E. R., 1954, *Political Systems of Highland Burma.* Athlone Press, London.
Lederman, R., 1986, *What Gifts Engender: Social Relations and Politics in Mendi, Highland Papua New Guinea.* Cambridge University Press, Cambridge.
Lehman, E. W., 1969, Toward a Macrosociology of Power. *American Sociological Review* 34: 453–465.
Leisner, G., & Leisner, V., 1943, Der Megalithgräber der Iberischen Halbinsel, I: der Süden. *Römisch-Germanische Forschungen* 17. Walter de Gruyter, Berlin.
Leonard, B. L., 1995, *Public Works in the Lower Chicama Valley, North Coast of Peru: AD 100–1470.* Paper presented at the 60th Annual Meetings of the Society for American Archaeology, Minneapolis.
Leonard, B. L., and Russell, G. S., 1994, *Coalescence and Transformation in the Early Intermediate Period: From Horizon to Horizon in the Chicama Valley, Peru.* Paper presented at the 59th Annual Meetings of the Society for American Archaeology, Anaheim.

REFERENCES

Leonard, R. D., and Jones, G. T., 1987, Elements of an Inclusive Evolutionary Model of Archaeology. *Journal of Anthropological Archaeology* 6(3): 199–219.
Lévi-Strauss, C., 1966, *The Savage Mind*. (English translation of *La Pensée Sauvage*, 1962). University of Chicago Press, Chicago.
Lévi-Strauss, C., 1982, *The Way of the Masks*. Translated from French by S. Modelski. University of Washington Press, Seattle.
Levy, J. E., 1992, *Orayvi Revisited: Social Stratification in an Egalitarian Society*. School of American Research Press, Santa Fe.
Lightfoot, D., 1993, The Cultural Ecology of Puebloan Pebble-Mulch Gardens. *Human Ecology* 21: 115–143.
Lightfoot, K., 1984, *Prehistoric Political Dynamics: A Case Study from the American Southwest*. Northern Illinois University Press, DeKalb.
Lipinski, E. (ed.), 1979, *State and Temple Economy in the Near East*. Departement Orientalistiek, Leuven.
Lizcano, R., Cámara, J. A., Riquelme, J. A., Cañabate, M. L., Sánchez, & Afonso, J. A., 1992, El Polideportivo de Martos: Producción Económica y Símbolos de Cohesión en un Asentamiento del Neolítico Final en las Campiñas del alto Guadalquivir. *Cuadernos de Prehistoria de la Universidad de Granada* 16–17: 5–101.
Lloyd, S., 1980, *Foundations in the Dust*. Thames and Hudson, New York.
Lofgren, L., and Turner, C., 1966, Household Size of Prehistoric Western Indians. *Southwestern Journal of Anthropology* 22: 117–132.
Louwe Kooijmans, L. P., 1998, Bronzezeitlichen Bauern in und um die niederländische Delta-Niederung. In *Mensch und Umwelt in der Bronzezeit Europas*, edited by B. Hänsel, pp. . Oetker-Voges Verlag, Kiel.
Lowie, R. H., 1920, *Primitive Society*. Horace Liveright, New York.
Lowie, R. H., 1927, *The Origin of the State*. Harcourt Brace, New York.
Lull, V., 1983, *La "Cultura" de El Argar*. Akal, Madrid.
Lull, V., 1984, A new assessment of Argaric economy and society. In *The Deya Conference of Prehistory: Early Settlement in the Western Mediterranean Islands and Their Peripheral Areas*, eduted by W. H. Waldren, R. Chapman, J. Lewthwaite, and R-C. Kennard, pp. 1197–1238. BAR International Series, Oxford.
Lull, V., 2000, El Argar: Death at Home. *Antiquity* 74: 581–590.
Lull, V., and Estévez, J., 1986, Propuesta Metodológica para el Estudio de las Necrópolis Argáricas. In *Homenaje a Luis Siret (1934–1984)*, pp. 441–452. Consejería de Cultura de la Junta de Andalucía, Sevilla.
Lull, V., and Risch, R., 1996, El Estado Argárico. *Verdolay* 7: 97–109.
Maekawa, K., 1980, Female Weavers and Their Children at Lagash: Pre-Sargonic and Ur III. *Acta Sumerologica* 2: 81–125.
Maekawa, K., 1987. Collective Labor Service in Girsu-Lagash: The Presargonic and Ur III Periods. In *Labor in the Ancient Near East*, edited by M. A. Powell, pp. 49–71. American Oriental Society, New Haven.
Maine, H., 1870, *Ancient Law*. John Murray, London.
Maldonado Cabrera, G., Molina González, F., Alcaraz Hernández, F., Cámara Serrano, J. A., Mérida González, V., & Ruiz Sánchez, V., 1992, El Papel Social del Megalitismo en el Sureste de la Península Ibérica: Las Comunidades Megalíticas del Pasillo de Tabernas. *Cuadernos de Prehistoria de la Universidad de Granada* 16–17: 167–190.
Mann, M., 1986, *The Sources of Social Power*. Volume 1: *A History of Power from the Beginning to AD 1760*. Cambridge University Press, Cambridge.
Manzanilla, L., 1997, Teotihuacan: Urban Archtype, Cosmic Model. In *Emergence and Change in Early Urban Societies*, edited by L. Manzanilla, pp. 109–131. Plenum Press, New York.

Manzanilla, L., 1997b, Early Urban Societies: Challenges and Perspectives. In *Emergence and Change in Early Urban Societies*, edited by L. Manzanilla, pp. 3–39. Plenum Press, New York.

Marcus, J., 1983, On the Nature of the Mesoamerican City. In *Prehistoric Settlement Patterns: Essays in Honor of Gordon R. Willey*, edited by E. Z. Vogt and R. M. Leventhal, pp. 195–242. University of New Mexico Press, Albuquerque, and Harvard University, Cambridge.

Marcus, J., 1992, *Mesoamerican Writing Systems: Propaganda, Myth and History in Four Ancient Civilizations*. Princeton University Press, Princeton.

Marcus, J., 1993, Ancient Maya Political Organization. In *Lowland Maya Civilization in the Eighth Century AD*, edited by J. A. Sabloff and J. S. Henderson, pp. 111–171. Dumbarton Oaks, Washington, DC.

Marcus, J., 1995, Where Is Lowland Maya Archaeology Headed? *Journal of Archaeological Research* 3: 3–53.

Marcus, J., 1998, The Peaks and Valleys of Archaic States: An Extension of the Dynamic Model. In *Archaic States*, edited by G. M. Feinman and J. Marcus, pp. 59–94. School of American Research Press, Santa Fe.

Marcus, J., and Flannery, K. V., 1996, *Zapotec Civilization: How Urban Society Evolved in Mexico's Oaxaca Valley*. Thames and Hudson, London.

Marinatos, N., 1995, Divine Kingship in Minoan Crete. In *The Role of the Ruler in Prehistoric Aegean*, edited by P. Rehak, pp 41–65. Aegaum 11, Belgium.

Marshall, F., 1990, Origins of Specialized Pastoral Production in East Africa. *American Anthropologist* 92(4): 873–894.

Martín, C., Fernández-Miranda, M., Fernández-Posse, M. D., and Gilman, A., 1993, The Bronze Age of La Mancha. *Antiquity* 67: 23–45.

Martin, S., and Grube, N., 1995, Maya Superstates. *Archaeology* 48(6): 41–46.

Martín de la Cruz, J. C., 1988, Mykenische Keramik aus Bronzezeitlichen Siedlungsschichten von Montoro am Guadalquivir. *Madrider Mitteilungen* 29: 77–92.

Martínez Navarrete, M. I., 1989, *Una Revisión Crítica de la Prehistoria Española: La Edad del Bronce como Paradigma*. Siglo Veintiuno de España, Madrid

Marx, K., 1964, *The Eighteenth Brumaire of Louis Bonaparte*. International Publishers, New York.

Maschner, H. D. G., 1991, The Emergence of Cultural Complexity. *Antiquity* 65(249): 924–934.

Maschner, H. D. G., 1995, Review of *A Complex Culture of the British Columbia Plateau: Traditional Stl'atl'imx Resource Use* (edited by B. Hayden). *American Antiquity* 60: 378–380.

Massey, S. A., 1986, Sociopolitical Change in the Upper Ica Valley, BC 400 to 400 AD Unpublished dissertation, University of California, Los Angeles.

Matheny, R., 1986, Investigations at El Mirador, Peten, Guatemala. *National Geographic Research* 2: 322–353.

Mathers, C., 1984, Beyond the Grave: The Context and Wider Implications of Mortuary Practice in Southeastern Spain. In *Papers in Iberian Archaeology*, edited by T. F. C. Blagg, R. J. F. Jones, and S. J. Keay, pp. 13–46. BAR International Series 193, Oxford.

Mathers, C., 1994, Goodbye to All That? Contrasting Patterns of Change in the Southeast Iberian Bronze Age c. 24/2200–600 BC. In *Development and Decline in the Mediterranean Bronze Age*, edited by C. Mathers and S. Stoddart, pp. 21–71. John Collins Publications, Sheffield.

Mathews, P., 1991, Classic Maya Emblem Glyphs. In *Classic Maya Political History: Hieroglyphic and Archaeological Evidence*, edited by T. P. Culbert, pp. 19–29. Cambridge University Press, Cambridge.

Matthiae, P., 1978, Preliminary Remarks on the Royal Palace of Ebla. *Syro-Mesopotamian*

REFERENCES

Studies 2/2: 13–41.
McAnany, P. A., 1993, The Economics of Social Power and Wealth among Eighth-Century Maya Households. In *Lowland Maya Civilization in the Eighth Century AD*, edited by J. A. Sabloff and J.S. Henderson, pp. 65–89. Dumbarton Oaks, Washington, DC.
McAnany, P. A., 1995, *Living with the Ancestors: Kinship and Kingship in Ancient Maya Society*. University of Texas Press, Austin.
McAnany, P. A., 1998, Ancestors and the Classic Maya Built Environment. In *Form and Meaning in Classic Maya Architecture*, edited by S. D. Houston, pp. 271–298. Dumbarton Oaks Research Library and Collection, Washington, DC.
McAnany, P. A., and López Varela, S. L., 1999, Re-Creating the Formative Maya Village of K'axob: Chronology, Ceramic Complexes, and Ancestors in Architectural Context. *Ancient Mesoamerica* 10: 147–168.
McAnany, P. A., Storey, R., and Lockard, A. K., 1999, Mortuary Ritual and Family Politics at Formative and Early Classic Kaxob, Belize. *Ancient Mesoamerica* 10: 129–146.
McCulloch, W. S., 1945, A heterarchy of values determined by the topology of nervous nets. *Bulletin of Mathematical Biophysics* 7: 89–93.
McIntosh, R. J., McIntosh, S. K., and Tainter, J., (eds.), 2001, *Archaeology and Social Memory*, Series in Historical Ecology, edited by W. Balee and C. Crumley, Columbia University Press.
Meltzer, D. J., Fowler, D. D., and Sabloff, J. A. (eds.), 1986, *American Archaeology Past and Future: A Celebration of the Society for American Archaeology*. Smithsonian Institution Press, Washington DC.
Michalowski, P., 1984, History as Charter: Some Observations on the Sumerian King List. *Journal of the American Oriental Society* 103: 237–248.
Midlarsky, M. I., 1999, *The Evolution of Inequality: War, State Survival, and Democracy in Comparative Perspective*. Stanford University Press, Stanford.
Mikkelsen, M., 1996, Bronzealderbosættelserne på Ås-højderyggen i Thy. In *Bronzealdererns bopladser i Midt-og Nordvestjylland*, edited by J. B. Bertelsen et al., pp. 110–123. Archaeolocial Museum of Viborg County, Viborg.
Milano, L., 1987, Food Rations at Ebla: A Preliminary Account on the Ration Lists Comong From the Ebla Palace Archive L.2712. *MARI* 5: 519–550.
Miller, D., 1985, Ideology and the Harappan Civilization. *Journal of Anthropological Archaeology* 4: 34–71.
Miller, M. E., and Houston, S. D., 1987, The Classic Maya Ball Game and its Architectural Setting. *RES, Anthropology and Aesthetics*, 14: 46–65.
Millon, C., 1973, Painting, Writings, and Polity in Teotihuacan, Mexico. *American Antiquity* 38: 294–314.
Millon, R., 1988, Where *Do* They All Come From? The Provenance of the Wagner Murals from Teotihuacan. In *Feathered Serpents and Flowering Trees*, edited by K. Berrin, pp. 78–113. Fine Arts Museums of San Francisco, San Francisco.
Millon, R., 1992, Teotihuacan Studies: From 1950 to 1990 and Beyond. In *Art, Ideology, and the City of Teotihuacan*, edited by J. C. Berlo, pp. 339–401. Dumbarton Oaks, Washington, DC.
Mills, B. J., 1997, *Gender, Craft Production, and Inequality in the American Southwest*. Paper presented at the School of American Research Advanced Seminar, "Sex Roles and Gender Hierarchies in the American Southwest," organized by P. L. Crown, Santa Fe.
Mingers, J., 1995, *Self-producing Systems: Implications and Applications of Autopoesis*. Plenum, New York.
Minsky, M., and S. Papert, 1972, *Artificial Intelligence Progress Report* (AI Memo 252). MIT Artificial Intelligence Laboratory, Cambridge.

Mitchell, W., 1991, *Peasants on the Edge: Crops, Cult, and Crisis in the Andes.* University of Texas Press, Austin.
Mithen, S., 1996, *The Prehistory of the Mind: The Cognitive Origins of Art, Religion and Science.* Thames and Hudson, London.
Molina González, F., 1983, *Prehistoria de Granada.* Editorial Don Quijote, Granada.
Molina González, F., 1988, El Calcolítico en la Península Ibérica: El Sudeste. *Rassegna di Archeologia,* 7: 255–262.
Molina González, F., Contreras Cortés, F., Ramos Millán, A., Mérida González, V., Ortiz Risco, F., and Ruiz Sánchez, V., 1986, Programa de Recuperación del Registro Arqueológico del Fortín 1 de Los Millares: Análisis Preliminar de la Organización del Espacio. *Arqueología Espacial* 8: 175–201.
Mommsen, T., 1968, *The Provinces of the Roman Empire: The European Provinces.* University of Chicago Press, Chicago.
Montero Ruiz, I., 1994, *El Origen de la Metalurgia en el Sureste Peninsular.* Instituto de Estudios Almerienses, Almeria.
Montero Ruiz, I., 1999, Sureste. In *Las Primeras Etapas Metalúrgicas en la Penísula Ibérica, Volume 2: Estudios Regionales,* edited by G. Delibes de Castro and I. Montero Ruiz, pp. 333–357. Instituto Universitario Ortega y Gasset, Madrid.
Montero Ruiz, I., and A. Ruiz Taboada, 1996, Enterramiento Colectivo y Metalurgia en el Yacimiento Neolitíco del Cerro Virtud (Cuevas de Almanzora, Almería). *Trabajos de Prehistoria* 53(2): 55–75.
Montero Ruiz, I., and Teneishvili, T. O., 1996, Estudio Actualizado de las Puntas de Jabalina del Dolmen de La Pastora (Valenciana de la Concepción, Sevilla). *Trabajos de Prehistoria* 53(1): pp. 73–90.
Moore, J., 1996a, The Archaeology of Plazas and the Proxemics of Ritual: Three Andean Traditions. *American Anthropologist* 98: 789–802.
Moore, J., 1996b, *Architecture and Power in the Prehispanic Andes.* Cambridge University Press, Cambridge.
Moorey, P. R. S., 1977, What Do We Know About the People Buried in the Royal Cemetery? *Expedition* 20(1): 24–40.
Moorey, P. R. S., 1978, *Kish Excavations 1923–1933.* Clarendon Press, Oxford.
Moreno Onorato, A., Contreras Cortés, F., and Cámara Serrano, J. A., 1992, Patrones de Asentamiento, Poblamiento y Dinámica Cultural en las Tierras Altas del Sureste Peninsular: El Pasillo Cúllar-Chirivel durante la Prehistoria Reciente. *Cuadernos de Prehistoria de la Universidad de Granada* 16–17: 191–245.
Morgan, L. H., 1877, *Ancient Society.* D. Appleton, New York.
Morrill, C., 1995, *The Executive Way: Conflict Management in Corporations.* University of Chicago Press, Chicago.
Moseley, M. E., 1975, *The Maritime Foundations of Andean Civilizations.* Cummings Publishing, Menlo Park.
Moseley, M. E., 1992, *The Inkas and Their Ancestors.* Thames and Hudson, London.
Mudar, K., 1982, Early Dynastic III Animal Utilization at Lagash: A Report on the Fauna of Tell al Hiba. *Journal of Near Eastern Studies* 41: 23–34.
Navarrete Enciso, M. S., 1976, *La Cultura de las Cuevas con Cerámica Decorada en Andalucía Oriental.* Universidad de Granada, Granada.
Neitzel, J. (ed.), 1999, *Great Towns and Regional Polities in the Prehistoric American Southwest and Southeast.* University of New Mexico Press, Albuquerque.
Netting, R. M., 1993, *Smallholders, Householders: Farm Families and the Ecology of Intensive, Sustainable Agriculture.* Stanford University Press, Stanford.
Nicholas, I., 1990, *Mayan Excavations I. Proto-Elamite Settlement at TUV.* University of Pennsylvania, University Museum Monograph 69, Philadelphia.

REFERENCES

Neiman, F. D., 1995, Stylistic Variation in Evolutionary Perspective. *American Antiquity* 60(1): 7–36.
Nissen, H., 1972, The City Wall of Uruk. In *Man, Settlement, and Urbanism*, edited by P. Ucko, R. Tringham, and G. W. Dimbleby. Duckworth, London.
Nissen, H., 1985, The Emergence of Writing in the Ancient Near East. *Interdisciplinary Science Reviews* 10(4): 349–361.
Nissen, H., 1986, The Archaic Texts from Uruk. *World Archaeology* 17: 317–334.
Nissen, H., Damerow, P., and Englund, R., 1993, *Archaic Bookkeeping. Early Writing and Techniques of Economic Administration in the Ancient Near East*. University of Chicago Press, Chicago.
Nocete, F., 1994, Space as Coercion: The Transition to the State in the Social Formations of La Campiña, Upper Guadalquivir Valley, Spain, ca. 1900–1600 BC. *Journal of Anthropological Archaeology* 13: 35–50.
O'Shea, J. M., 1996, *Villagers of the Maros. A Portrait of an Early Bronze Age Society*. Plenum Press, New York and London.
Oliver, D. L., 1970, A Leader in Action. In *Cultures of the Pacific: Selected Readings*, edited by T. G. Harding and B. J. Wallace, pp. 216–231. Free Press, New York.
Olmsted, G. S., 1994, The Gods of the Celts and the Indo-Europeans. *Archaeolongua*. Innsbrucker Beiträge zur Kultruwissenschaft.
Oppenheim, A. L., 1977, *Ancient Mesopotamia*. University of Chicago Press, Chicago.
Ottenberg, S., 1982, Illusion, Communication, and Psychology in West Africa Masquerades. *Ethos* 10: 149–185.
Otterbein, K. F., 1985, *The Evolution of War: A Cross-Cultural Study*. HRAF Press, New Haven.
Ortiz, P., and del C. Rodriguez, M., 1994, Los Espacios Sagrados Olmecas: El Manatí, un Caso Especial. In *Los Olmecas en Mesoamérica*, edited by J. E. Clark, pp. 69–91. El Equilibrista, México, D. F.
Owen, B., 1993, *Early Ceramic Settlement in the Coastal Osmore Valley, Peru: Preliminary Report*. Paper presented at the 33rd Annual Meeting of the Institute of Andean Studies, Berkeley.
Owen, B., and M. A. Norconk, 1987, Analysis of the Human Burials, 1977–1983 Field Seasons: Demographic Profiles and Burial Practices (Appendix 1). In *Archaeological Field Research in the Upper Mantaro, Peru, 1982–1983. Investigations of Inka Expansion and Exchange*. Monograph XXVII, Institute of Archaeology, edited by T. Earle, T. N. D'Altroy, C. Hastorf, C. J. Scott, C. L. Costin, G. S. Russell, and E. C. Sandefeur, pp. 107–123. University of California, Los Angeles.
Palaima, T. G., 1995, The Nature of the Mycenaean Wanax. Non-Indo-European Origins and Priestly Functions. In *The Role of the Ruler in Prehistoric Aegean*, edited by P. Rehak, Aegaum II. Annales d'archeologie egeene de l'Universte de Liege et UT-PASP, Liege.
Pantaleón-Cano, J., Roure, J. M., Yll, E. I., & Pérez Obiol, R., 1996, Dinámica del Paisaje Vegetal Durante el Neolítico en la Vertiente Mediterránea de la Península Ibèrica e Islas Baleares. In *Rubricatum (Actes, I Congrés del Neolític a la Península Ibàrica, Formació i implantació de les Comunitats Agrícoles, Gavà-Bellaterra, 27, 28 i 29 de Març de 1995)*, Volume 1(1), pp. 29–34. Museu de Gavà, Gavà.
Pantaleón-Cano, J., Yll, R., and Roure, J. M., 1999, Evolución del Paisaje Vegetal en el Sudeste de la Península Ibérica Durante el Holoceno a Partir del Análisis Polínico. In *Actes del II Congrés del Neolític a la Península Ibérica, Universitat de Valencia, 7-9 d=Abril, 1999*, edited by J. Bernabeu Aubán and T. Orozco Köhler, pp. 17–23. Universitat de Valencia, Valencia.
Parsons, M. H., 1970, Preceramic Subsistence on the Peruvian Coast. *American Antiquity* 35: 292–304.
Pasztory, E., 1978, Artistic Traditions of the Middle Classic Period. In *Middle Classic*

Mesoamerica: AD 400–700, edited by E. Pasztory, pp. 108–142. Columbia University Press, New York.

Pasztory, E., 1988, A Reinterpretation of Teotihuacan and Its Mural Painting Tradition. In *Feathered Serpents and Flowering Trees*, edited by K. Berrin, pp. 45–77. Fine Arts Museums of San Francisco, San Francisco.

Pasztory, E., 1992, Abstraction and the Rise of a Utopian State at Teotihuacan. In *Art, Ideology, and the City of Teotihuacan*, edited by J.C. Berlo, pp. 281–320. Dumbarton Oaks, Washington, D.C.

Pasztory, E., 1997, *Teotihuacan: An Experiment in Living*. University of Oklahoma Press, Norman.

Patterson, T. C., 1985, The Huaca la Florida, Rimac Valley, Peru. In *Early Ceremonial Architecture in the Andes*, edited by C. B. Donnan, pp. 59–70. Dumbarton Oaks, Washington, DC.

Patterson, T. C., and Gailey, C. W., eds., 1987, *Power Relations and State Formation*. American Anthropological Association, Washington, DC.

Pauketat, T. R., 1996, The Foundations of Inequality within a Simulated Shan Community. *Journal of Anthropological Archaeology* 15(3): 219–236.

Pauketat, T. R., and Emerson, T. E., 1999, Represenations of Hegemony as Community at Cahokia. In *Material Symbols: Culture and Economy in Prehistory*, edited by J. E. Robb, pp. 302–313. Occasional Paper No. 26, Center for Archaeological Investigations, Southern Illinois University, Carbondale.

Paulinyi, Z., 1981, Capitals in Pre-Aztec Central Mexico. *Acta Orientalia Academiae Hungaricae* 35: 315–350.

Payne, S., 1973, Kill-Off Patterns in Sheep and Goats: The Mandibles from Asvan Kale. *Anatolian Studies* 23: 281–303.

Paynter, R., 1989, The Archaeology of Equality and Inequality. *Annual Review of Anthropology* 1989, 18: 369–399.

Pendergast, D. M., 1982, *Excavations at Altun Ha, Belize, 1964–1970, Volume 2*. Royal Ontario Museum, Toronto.

Pellicer, M., and Acosta, P., 1986, Neolítico y Calcolítico de la Cueva de Nerja. In *La Prehistoria de la Cueva de Nerja (Málaga)*, edited by J. F. Jordá Pardo. Patronato la Cueva de Nerja.

Peters, J., and von den Driesch, A., 1990, Archäozoologische Untersuchung der Tierreste aus der kupferzeitlichen Siedlung von Los Millares (Prov. Almería). In *Studien über frühe Tierknochenfunde von der Iberischen Halbinsel*, Volume 12, pp. 51–115. Institut für Palaeoanatomie, Domestikationsforschung und Geschichte der Tiermedizin der Universität München, Munich.

Pinnock, F., 1988, Observations on the Trade of Lapis Lazuli in the 3rd Millennium BC. In *Wirtschaft und Gesellschaft von Ebla*, edited by H. Waetzoldt and H. Hauptmann, pp. 107–110. Heidelberger Orientverlag, Heidelberg.

Plog, S., 1997, *Ancient Peoples of the American Southwest*. Thames and Hudson, New York.

Polanyi, K., 1944, *Great Transformations*. Beacon, New York.

Pollock, S., 1991a, Of Priestesses, Princes, and Poor Relations: The Dead in the Royal Cemetery of Ur. *Cambridge Archaeological Journal* 1(2): 171–189.

Pollock, S., 1991b, Women in a Men's World: Images of Sumerian Women. In *Engendering Archaeology: Women and Prehistory*, edited by J. Gero and M. Conkey, pp. 366–387. Blackwell, Oxford.

Pollock, S., Pope, M., and Coursey, C., 1996, Household Production at the Uruk Mound, Abu Salabikh, Iraq. *American Journal of Archaeology* 100: 683–698.

Pons, A., and Reille, M., 1988, The Holocene and Upper Pleistocene Pollen Record from Padul (Granada, Spain): A New Study. *Palaeogeography, Palaeoclimatology, Palaeoecology* 66: 243–263.

REFERENCES

Possehl, G. L., 1990, Revolution in the Urban Revolution: The Emergence of Indus Urbanization. *Annual Review of Anthropology* 19: 261–282.
Possehl, G. L., 1998, Sociocultural Complexity without the State: The Indus Civilization. In *Archaic States*, edited by G. M. Feinman and J. Marcus, pp. 261–291. School of American Research, Santa Fe.
Postgate, J. N., 1975, Some Old Babylonian Shepherds and Their Flocks. *Journal of Semitic Studies*: 1–21.
Postgate, J. N., 1983, Changing Subsistence Priorities and Early Settlement Patterns on the North Coast of Peru. *Journal of Ethnobiology* 3: 15–38.
Postgate, J. N., 1986, The Transition from Uruk to Early Dynastic: Continuities and Discontinuities in the Record of Settlement. In *Gamdat Nasr: Period or Regional Style?* edited by U. Finkbeiner and W. Röllig, pp. 90–106. TAVO Reihe B62, Wisbeiden.
Postgate, J. N., 1992, *Early Mesopotamia: Society and Economy at the Dawn of History*. Routledge, London.
Postgate, J. N., 1994, How Many Sumerians per Hectare? Probing the Anatomy of an Early City. *Cambridge Archaeological Journal* 4: 47–65.
Poulsen, J., 1983, Nogle reflektioner omkring Vognserup Engefundet. In *Struktur och Ferandring i Bronsalderns Samhalle*. Rapport frun det Tredje Nordiska Symposiet for Bronsaldersforskning i Lund 32-25 April 1982. University of Lund, Institute of Archaeological Report Series 17, Lund.
Powell, M., 1985, Salt, Seed, and Yields in Sumerian Agriculture. A Critique of the Theory of Progressive Salinization. *Zeitschrift für Assyriologie* 75: 7–38.
Pozorski, S., 1976, *Prehistoric Subsistence Patterns and Site Economics in the Moche Valley, Peru*. Unpublished Ph.D. dissertation, University of Texas, Austin.
Pozorski, S., 1986, Recent Excavations at Pampa de Las Llamas-Moxeke, a Complex Initial Period Site in Peru. *Journal of Field Archaeology* 13: 381–401.
Pozorski, S., 1987, *Early Settlement and Subsistence in the Casma Valley Peru*. University of Iowa Press, Iowa City.
Pozorski, S., 1992, Early Civilization in the Casma Valley, Peru. *Antiquity* 270(6): 66–72.
Pozorski, S., and Pozorski, T., 1979, Alto Salaverry: A Peruvian Coastal Cotton Preceramic Site. *Annuals of the Carnegie Museum* 49: 337–375.
Pozorski, T., 1976, *Caballo Muerto: A Complex of Early Ceramic Sites in the Moche Valley, Peru*. Unpublished Ph.D. dissertation. University of Texas, Austin.
Pozorski, T., 1980, The Early Horizon Site of Huaca de los Reyes: Societal Implications. *American Antiquity* 45: 100–110.
Pozorski, T., 1982, Early Social Stratification and Subsistence Systems: The Caballo Muerto Complex. In *Chan Chan: Andean Desert City*, edited by M. E. Moseley and K. C. Day, pp. 225–253. University of New Mexico Press, Albuquerque.
Pozorski, T., and Pozorski, S., 1990, Huyanuná, a Late Cotton Preceramic Site on the North Coast of Peru. *Field Journal of Archaeology* 17: 17–26.
Price, B., 1977, Shifts of Production and Organization: A Cluster Interaction Model. *Current Anthropology* 18(2): 209–234.
Price, T. D., & Feinman, G. M. (eds.), 1995, *Foundations of Social Inequality*. Plenum Press, New York.
Quilter, J., 1985, Architecture and Chronology at El Paraíso, Peru. *Journal of Field Archaeology* 12: 279–297.
Quin, W. H., and Neal, V. T., 1987, El Niño Occurrences Over the Past Four and a Half Centuries. *Journal of Geophysical Research* 92: 14,449–461.
Radcliffe-Brown, A. R., 1952, *Structure and Function in Primitive Society*. Free Press, New York.
Ramos Millán, A., 1981, Interpretaciones secuenciales y culturales de la Edad del Cobre en la

zona meridional de la Península Ibérica: La alternativa del materialismo cultural. *Cuadernos de Prehistoria de la Universidad de Granada* 6: 242–256.

Ramos Millán, A., 1997, Flint Political Economy in a Tribal Society: A Material-Culture Study in the El Malagón Settlement (Iberian Southeast). In *Siliceous Rocks and Culture*, edited by A. Ramos Millán and M. A. Bustillo, pp. 671–711. Universidad de Granada, Granada.

Randsborg, K., 1993, Kivik. Archaeology and Iconography. *Acta Archaeologica* 64(1).

Randsborg, K., 1995, *Hjortspring. Warfare and Sacrifice in Early Europe*. Aarhus University Press, Aarhus.

Rapoport, A., 1994, Spatial Organization and the Built Environment. In *Companion Encyclopedia of Anthropology: Humanity, Culture and Social Life*, edited by T. Ingold, pp. 460–502. Routledge, London.

Rathje, W. L., 1973, Classic Maya Development and Denouement: A Recent Design. In *The Classic Maya Collapse*, edited by T. P. Culbert, pp. 405–454. University of New Mexico Press, Albuquerque.

Rathje, W. L., 1980, Socio-political Implications of Lowland Maya Burials: Methodology and Tentative Hypotheses. *World Archaeology* 1: 359–374.

Ravines, R., 1985, Early Monumental Architecture of the Jequetepeque Valley, Peru. In *Early Ceremonial Architecture in the Andes*, edited by C. B. Donnan, pp. 209–226. Dumbarton Oaks, Washington, DC.

Redding, R., 1981, *Decision Making in Subsistence Herding of Sheep and Goats in the Middle East*. Ph.D. dissertation, University of Michigan, Ann Arbor.

Redman, C., 1978, *The Rise of Civilization*. W.H. Freeman, San Francisco.

Reed, L. W., 1990, X-Ray Diffraction Analysis of Glaze Painted Ceramics from the Northern Rio Grande Region, New Mexico: Implications of Glazeware Production and Exchange. In, *Economy and Polity in Late Rio Grande Prehistory*, edited by S. Upham and B. D. Staley, pp. 90–149. University Museum, New Mexico State University, Occasional Papers 16, Las Cruces.

Reed, P. F., 1990, A Spatial Analysis of the Northern Rio Grande Region, New Mexico. In *Economy and Polity in Late Rio Grande Prehistory*, edited by S. Upham and B. D. Staley, pp. 90–149. University Museum, New Mexico State University, Occasional Papers 16, Las Cruces.

Reents-Budet, D., 1994, *Painting the Maya Universe: Royal Ceramics of the Classic Period*. Duke University Press, Durham.

Reid, T. R., 1997, The Power and the Glory of the Roman Empire. *National Geographic* 192: 2–41.

Rehak, P., 1995, Enthroned Figures in Aegean Art and the Function of the Mycenaean Megaron. In *The Role of the Ruler in the Prehistoric Aegean*, edited by P. Rehak, pp. . Aegaeum II. Annales d'Archeologie Eegeene de l'Universite de Liege et UT-PASP, Liege.

Renfrew, C., 1967, Colonialism and Megalithismus. *Antiquity* 41: 276–288.

Renfrew, C., 1973a, Monuments, Mobilisation, and Social Organisation in Neolithic Wessex. In *The Explanation of Culture Change: Models in Prehistory*, edited by C. Renfrew, pp. 539–558. Duckworth, London.

Renfrew, C., 1973b, *Before Civilization: The Radiocarbon Revolution and Prehistoric Europe*. Jonathan Cape, London.

Renfrew, C., 1974, Beyond a Subsistence Economy: The Evolution of Social Organisation in Prehistoric Europe. In *Reconstructing Complex Societies: An Archaeological Colloquium*, edited by C. B. Moore, pp. 69–95. Supplement to the Bulletin of the American Schools of Oriental Research 20. Ann Arbor.

Renfrew, C., and Cherry, J. F., (eds.), 1986, *Peer Polity Interaction and Socio-political Change*. Cambridge University Press, Cambridge.

REFERENCES

Renfrew, J. M., 1973, *Palaeoethnobotany*. Methuen, London.
Restall, M., 1997, *The Maya World: Yucatec Culture and Society, 1550–1850*. Stanford University Press, Stanford.
Rice, D. S., 1986, The Petén Postclassic: A Settlement Perspective. In *Late Lowland Maya Civilization: Classic to Postclassic*, edited by J. A. Sabloff and E. W. Andrews V, pp. 301–344. University of New Mexico Press, Albuquerque.
Risch, R., 1998, Análisis Paleoeconómico y Medios de Producción Líticos: El Caso de Fuente Álamo. In *Minerales y Metales en la Prehistoria Reciente: Algunos Testimonios de su Explotación y Laboreo en la Península Ibérica*, edited by G. Delibes de Castro, pp. 105–154. Studia Archaeologica 88, Universidad de Valladolid, Valladolid.
Robertson, M. G., 1983, *The Sculpture of Palenque, Volume I, The Temple of the Inscriptions*. Princeton University Press, Princeton.
Robin, C., 1989, *Preclassic Maya Burials at Cuello, Belize*. BAR International Series 480, Oxford.
Rodríguez-Ariza, M. O., 1997, Contrastación de la Vegetación Calcolítica y Actual en la Cuenca del Ándarax a Partir de la Antracología. In *Anuario Arqueológico de Andalucía*, Volume 2: *Actividades Sistemáticas*, pp. 14–23. Consejería de Cultura de la Junta de Andalucía, Sevilla.
Rodríguez-Ariza, M. O., & Esquivel, J. A., 1990, Una Aplicación del Análisis de Correspondencias en la Valoración del Antracoanálisis de Los Millares. *Cuadernos de Prehistoria de la Universidad de Granada* 14–15: 81–108.
Rodríguez-Ariza, M. O., & Vernet, J. L., 1992, Premiers résultats paléoécologiques de létablissement chalcolithique de Los Millares (Santa Fe de Mondújar, Almería, Espagne), d'après l'analyse anthracologique de l'établissement. In *IInd Deya International Conference of Prehistory, Volume 1: Archaeological Techniques and Technology*, edited by W. H. Waldren, J. A. Ensenyat, and R. C. Kennard, pp. 3–16. BAR International Series 573, Oxford.
Roscoe, P. B., 1993, Practice and Political Centralization. *Current Anthropology* 34: 111–140.
Rosenswig, R. M., 1998, *A Comparison of Early and Middle Formative Political Development in the Soconusco and Valley of Oaxaca: Settlement, Mortuary, and Architectural Patterns*. Unpublished Masters thesis, Department of Anthropology and Sociology, University of British Columbia.
Rothman, M., 1994, Palace and Private Agricultural Decision-Making in the Early 2nd Millennium BC City-State of Larsa, Iraq. In *The Economic Anthropology of the State*, edited by E. Brumfiel, pp. 149–166. Monographs in Economic Anthropology 11. University Press of America, Lanham.
Rousseau, J-J., 1947, The Social Contract. In *Social Contract*, edited by E. Barker. Oxford University Press, New York.
Roys, R. L., 1967, *The Book of Chilam Balam of Chumayel*. (Originally published in 1933 in CIW series). University of Oklahoma Press, Norman.
Ruiz, M., Risch, R., González Marcén, P., Castro, P., Lull, V., and Chapman, R., 1992, Environmental Exploitation and Social Structure in Prehistoric Southeast Spain. *Journal of Mediterranean Archaeology* 5: 3—38.
Ruiz Taboada, A., and Montero Ruiz, I., 1999, The Oldest Metallurgy in Western Europe. *Antiquity* 73: 897–903.
Russell, G., 1988, *The Effect of Inka Administrative Policy on the Domestic Economy of the Wanka, Peru: The Production and Use of Stone Tools*. Ph.D. dissertation, Department of Anthropology, UCLA.
Ruz Lhuiller, A., 1973, *El Templo de las Inscripciones Palenque*. Instituto Nacional de Antropología e Historia, Mexico.

Sabloff, J. A., and Rathje, W. R., 1975, The Rise of the Maya Merchant Class. *Scientific American* 233(4): 72–82.
Sáez, L., and Martínez, G., 1981, El Yacimiento Neolítico al Aire Libre de La Molaina (Pinos Puente, Granada). *Cuadernos de Prehistoria de la Universidad de Granada* 6: 17–34.
Sahlins, M., 1972, *Stone Age Economics*. Aldine, Chicago.
Sahlins, M., and Service, E . R. (eds.), 1960, *Evolution and Culture*. University of Michigan Press, Ann Arbor.
Sampson, C. G., 1988, *Stylistic Boundaries among Mobile Hunter-Foragers*. Smithsonian Institution Press, Washington, DC.
Sanders, W. T., 1981, Classic Maya Settlement Patterns and Ethnographic Analogy. In *Lowland Maya Settlement Patterns*, edited by W. Ashmore, pp. 351–369. University of New Mexico Press, Albuquerque.
Sanders, W. T., and Price, B. J., 1968, *Mesoamerica: The Evolution of a Civilization*. Random House, New York.
Sandford, S., 1982, Pastoral Strategies and Desertification: Opportunism and Conservatism in Arid Lands. In *Desertification and Development: Dryland Ecology in Social Perspective*, edited by B. Spooner and H. S. Mann, pp. 61–80. Academic Press, London.
Sandford, S., 1983, *Management of Pastoral Development in the Third World*. John Wiley and Sons, Chichester.
Sanmartin, J., 1993, Sheep and Goats in the Akkadian Economic Texts from Ugarit. *Bulletin on Sumerian Agriculture* 7: 199–207.
Schaafsma, C. F., 1995, The Chronology of Las Madres Pueblo (LA 25). *Papers of the Archaeological Society of New Mexico* 21: 155–164.
Schauer, P., 1984, Überregionale Gemeinsamkeiten bei Waffengräbern der Ausgehenden Bronzezeit und Älteren Urnenfelderzeit des Voralpenraumes. *Jahrbuch des Römisch-Germanischen Zentralmuseums*, 31:209–235. Mainz.
Schauer, P., 1990, Schutz- und Angriffswaffen bronzezeitlicher Krieger im Spiegel ausgewählter Grabfunde Mitteleuropas. In *Beitrage zur Geschichte und Kultur der Mitteleuropaischen Bronzezeit*, edited by V. Furmanek and F. Horst, Nitra, Berlin.
Schele, L., 1992, The Founders of Lineages at Copan and Other Maya Sites. *Ancient Mesoamerica* 3(1): 135–144.
Schele, L., and Miller, M. E., 1986, *The Blood of Kings: Dynasty and Ritual in Maya Art*. Kimball Art Museum, Fort Worth.
Schieve, W. C., and Allen, P. M. (eds.), 1982, *Self-Organization and Dissipative Structures: Applications in the Physical and Social Sciences*. University of Texas Press, Austin.
Schmandt-Besserat, D., 1993, Images of Enship. In *Between the Rivers and Over the Mountains*, edited by M. Frangipane, H. Hauptmann, M. Liverani , P. Matthiae, and M. Mellink, pp. 201–219. Universita di Roma "La Sapienza," Rome.
Schneider, P., Schneider, J., and Hansen, E., 1972, Modernization and Development: The Role of Regional Elites and Noncorporate Groups in the European Mediterranean: *Comparative Studies in Society and History* 14: 328–350.
Schortman, E. M., and Urban, P. A. (eds.), 1992, *Resources, Power, and Interregional Interaction*. Plenum Press, New York.
Schreiber, K. J., 1992, *Wari Imperialism in Middle Horizon Peru*. Anthropological Papers of the Museum of Anthropology, No. 87. University of Michigan, Ann Arbor.
Schreiber, K. J., 1999, Settlement Surveys and the Archaeological Study of Complex Societies: Examples from Ayacucho and Nasca. In *Settlement Pattern Studies in the Americas: Fifty Years Since Viru*, edited by B. R. Billman and G. M. Feinman, pp. 160–171. Smithsonian Institution Press, Washington, DC.
Schubart, H., Arteaga, O., & Pingel, V., 1985, Fuente Alamo: Informe Preliminar Sobre la Excavación de 1985 en el Poblado de la Edad del Bronce. *Empúries* 47: 70–107.

REFERENCES

Schüle, W., 1967, Feldbewässerung in Alt-Europa. *Madrider Mitteilungen* 8: 79–99.
Schüle, W., 1986, El Cerro de la Virgen de la Cabeza (Orce, Granada): Consideraciones Sobre su Marco Ecológico y Cultural. In *Homenaje a Luis Siret (1934–1984)*, edited by O. Arteaga, pp. 208–220. Consejería de Cultura de la Junta de Andalucía, Sevilla.
Schwartz, G., 1988, *A Ceramic Chronology from Tell Leilan*. Yale University Press, New Haven.
Schwartz, G., and Curvers, H., 1992, Tell al-Raqa'i 1989 and 1990: Further Investigations at a Small Rural Site of Early Urban North Mesopotamia. *American Journal of Archaeology* 96: 397–419.
Scott, G. P. (ed.), 1991, *Time, Rhythms, and Chaos in the New Dialogue with Nature*. University of Iowa Press, Ames.
Sebastian, L., 1992, *The Chaco Anasazi: Sociopolitical Evolution in the Prehistoric Southwest*. Cambridge University Press, Cambridge.
Sempowski, M. L., 1987, Differential Mortuary Treatment: Its Implications for Social Status at Three Residential Compounds in Teotihuacan, Mexico. In *Teotihuacan: Nuevos Datos, Nuevas Síntesis, Nuevos Problemas*, edited by E. McClung de Tapia and E. C. Rattray, pp. 115–131. Universidad Nacional Autónoma de México, DF.
Sempowski, M. L., 1992. Economic and Social Implications of Variations in Mortuary Practices at Teotihuacan. In *Art, Ideology, and the City of Teotihuacan*, edited by J. C. Berlo, pp. 27–58. Dumbarton Oaks, Washington, DC.
Sempowski, M. L., and Spence, M. W., 1994, *Mortuary Practices and Skeletal Remains at Teotihuacan*. University of Utah Press, Salt Lake City.
Senior, L., and Weiss, H., 1992, Tell Leilan "Sila Bowls" and the Akkadian Reorganization of Subarian Agricultural Production. *Orient Express* 2: 16–21.
Service, E. R., 1962, *Primitive Social Organization*. Random House, New York.
Service, E. R., 1971a, *Primitive Social Organization: An Evolutionary Perspective*. 2nd ed. Random House, New York.
Service, E. R., 1971b, *Profiles in Ethnology*. Harper and Row, New York.
Service, E. R., 1975, *The Origins of the State and Civilization: The Process of Cultural Evolution*. Norton Press, New York.
Shafer, H. J., and Hester, T. R., 1991, Lithic Craft Specialization and Product Distribution at the Maya site of Colha, Belize. *World Archaeology* 23(1): 79–97.
Shaffer, J. G., 1982, Harappan Culture: A Reconsideration. In *Harappan Civilization: A Contemporary Perspective*, edited by G. L. Possehl, pp. 41–50. Oxford and IBH, and the American Institute of Indian Studies, Delhi.
Shanks, M., 1992, Style and the Design of a Perfume Jar from an Archaic Greek City State. *Journal of European Archaeology* 1:77–106.
Sharer, R. J., 1992, The Preclassic Origins of Lowland Maya States. In *New Theories on the Ancient Maya*, edited by E. C. Danien and R. J. Sharer, pp. 131–136. The University Museum, University of Pennsylvania, Philadelphia.
Sharer, R. J., 1994, *The Ancient Maya*. 5th edition. Stanford University Press, Stanford.
Sharer, R. J., and Sedat, D. W., 1987, *Archaeological Investigations in the Northern Highlands, Guatemala: Interaction and the Development of Maya Civilization*. University Museum Monograph 59. The University Museum, University of Pennsylvania, Philadelphia.
Shennan, S., 1982, Exchange and Ranking: The Role of Amber in the Earlier Bronze Age. In *Ranking, Resources and Exchange*, edited by C. Renfrew and S. Shennan, pp. 33–45. Cambridge University Press, Cambridge.
Shennan, S., 1989, *Archaeological Approaches to Cultural Identity*. Unwin Hyman, Ltd., London.
Shepard, A. O., 1942, *Rio Grande Glaze Paint Ware*. Contributions to American Anthropology and History 39. Publication 528. Carnegie Institution of Washington, Washington, DC.

Sherratt, A., 1981, Plough and Pastoralism: Aspects of the Secondary Products Revolution. In *Pattern of the Past: Studies in Honour of David Clarke*, edited by I. Hodder, G. Isaac, and N. Hammond, pp. 261–305. Cambridge University Press, Cambridge.

Shimada, I., Elera, C. G., and Shimada, M., 1982, Excavaciones Efectuadas en el Centro Ceremonial de Huaca Lucia-Calliope del Horizonte Temprano, Batán Grande, Costa Norte del Peru: 1979–81. *Archeológicas* 19: 109–201.

Silverman, H., 1993, *Cahuachi in the Ancient Nasca World*. University of Iowa Press, Iowa City.

Silverman, H., 1994, Paracas in Nasca: New Data on the Early Horizon Occupation of the Rio Grande de Nasca Drainage, Peru. *Latin American Antiquity* 5: 359–382.

Simon, H., 1959, Theories of Decision Making in Economics and Behavioral Science. *American Economic Review* 49(3): 253–283.

Sinopoli, C. M., 1994, The Archaeology of Empires. *Annual Review of Anthropology* 23: 159–180.

Siret, H., and Siret, L., 1887, *Les Premiers Ages du Metal dans le Sud-Est de l'Espagne*. Antwerp.

Siret, L., 1893, L'Espagne Préhistorique. *Revue des Questions Scientifiques* 4: 529–582.

Smith, A. B., 1992, Origins and Spread of Pastoralism in Africa. *Annual Review of Anthropology* 21: 125–141.

Snow, D. H., 1973, Prehistoric Southwestern Turquoise Industry. *El Palacio* 79(1): 33–51.

Snow, D. H., 1981, Protohistoric Rio Grande Economics: A review of trends. In *The Protohistoric in the North American Southwest, AD 1450–1700*, edited by D. R. Wilcox and W. B. Masse, pp. 354–377. Arizona State University Research Papers 24, Tempe.

Soltis, J., Boyd, R., and Richerson, P. J., 1995, Can Group-functional Behaviors Evolve by Cultural Group Selection?: An Empirical Test. *Current Anthropology* 36(3): 473–494.

Sørensen, M. L. S., 1997, Reading Dress: The Construction of Social Categories and Identitites in Bronze Age Europe. *Journal of European Archaeology*, vol.5, No.1.

Spencer, C., 1987, Rethinking the Chiefdom. In *Chiefdoms in the Americas*, edited by R. Drennan and C. Uribe, pp. 369–390. University Press of America, Lanham.

Spencer, C. S., 1993, Human Agency, Biased Transmission, and the Cultural Evolution of Chiefly Societies. *Journal of Anthropological Archaeology* 12: 41–74.

Spielmann, K., 1994, Clustered Confederacies: Sociopolitical Organization in the Protohistoric Rio Grande. In *The Ancient Southwestern Community*, edited by W. H. Wills and R. D. Leonard, pp. 45–54. University of New Mexico Press, Albuquerque.

Stanish, C., 1999. Settlement Shifts and Political Ranking in the Lake Titicaca Basin, Peru. In *Settlement Pattern Studies in the Americas: Fifty Years since Viru*, edited by B. R. Billman and G. M. Feinman, pp. 116–130. Smithsonian Institution Press, Washington, DC.

Stark, B. L., and Arnold, P. J., (eds.), 1997, *Olmec to Aztec: Settlement Patterns in the Ancient Gulf Lowlands*. University of Arizona Press, Tucson.

Stark, D., n.d., *Heterarchy: Asset Ambiguity, Organizational Innovation, and the Postsocialist Firm*. Paper prepared for the Thematic Session on the Firm in the 21st Century. Annual Meetings of the American Sociological Association, New York, August 16–20, 1996.

Startin, D. W. A., 1982, Prehistoric earthmoving. In *Settlement Patterns in the Oxford Region*, edited by H. J. Chase and A. Whittle, pp. 153–15-6. Oxford University Press, Oxford.

Stein, G., 1987, Regional Economic Integration in Early State Societies: Third Millennium BC Pastoral Production at Gritille, Southeast Turkey. *Paléorient* 13(2): 101–111.

Stein, G., 1988, *Pastoral Production in Complex Societies: Mid-Late Third Millennium BC and Medieval Faunal Remains from Gritille Höyük in the Karababa Basin, southeast Turkey*. PhD Dissertation, University of Pennsylvania.

Stein, G., 1994a, Introduction Part II. The Organizational Dynamics of Complexity in Greater Mesopotamia. In *Chiefdoms and Early States in the Near East: The Organizational Dy-

REFERENCES

namics of Complexity, edited by G. Stein and M. Rothman, pp. 11–22. Monagraphs in World Prehistory 18. Prehistory Press, Madison.

Stein, G., 1994b, Economy, Ritual, and Power in 'Ubaid Mesopotamia. In *Chiefdoms and Early States in the Near East: The Organizational Dynamics of Complexity,* edited by G. Stein and M. S. Rothman, pp. 35–46. Prehistory Press, Madison.

Stein, G., 1996, Producers, Patrons, and Prestige: Craft Specialists and Emergent Elites in Mesopotamia from 5500–3100 BC. In *Craft Specialization and Social Evolution: In Memory of V. Gordon Childe,* edited by B. Wailes, pp. 25–38. University of Pennsylvania, University Museum. Museum Monograph 94, Philadelphia.

Stein, G., 1998, Heterogeneity, Power, and Political Economy: Some Current Research Issues in the Archaeology of Old World Complex Societies. *Journal of Archaeological Research* 6(1): 1–44.

Stein, G., and Blackman, M. J., 1993, The Organizational Context of Specialized Craft Production in Early Mesopotamian States. *Research in Economic Anthropology* 14: 29–59.

Stein, G., and Rothman, M. S. (eds.), 1994, *Chiefdoms and Early States in the Near East: The Organizational Dynamics of Complexity.* Prehistory Press, Madison.

Stein, G., and Wattenmaker, P., 1990, The 1987 Tell Leilan Regional Survey: Preliminary Report. In *Economy and Settlement in the Near East: Analyses of Ancient Sites and Materials. MASCA Research Papers in Science and Archaeology* 7 (supplement), edited by N. Miller, pp. 8–18. University of Pennsylvania, University Museum, Philadelphia.

Steiner, C. B., 1990, Body Personal and Body Politic: Adornment and Leadership in Cross-Cultural Perspective. *Anthropos* 85: 431–445.

Steponaitis, V. P., 1981, Settlement Hierarchies and Political complexity in Nonmarket Societies: The Formative Period in the Valley of Mexico. *American Anthropologist,* 83: 320–363.

Steward, J., 1951, Levels of Sociocultural Integration: An Organizational Concept. *Southwestern Journal of Anthropology* 7: 374–390.

Steward, J., 1955, *Theory of Culture Change.* University of Illinois Press, Urbana.

Stika, H. P., 1988, Botanische Untersuchungen in der bronzezeitlichen Höhensiedlung Fuente Álamo. *Madrider Mitteilungen* 29: 21–76.

Stika, H. P., 1998, Landwirtschaft, Klima und Umwelt zur Bronzezeit im semiariden Becken von Vera, Prov. Almería, Südostspanien. In *Mensch und Umwelt in der Bronzezeits Europas,* edited by B. Hänsel, pp. 111–115. Oetker-Voges Verlag, Kiel.

Stika, H. P., and Jurich, B., 1998, Pflanzenreste aus der Probegrabung 1991 im bronzezeitlichen Siedlundsplatz El Argar, Prov. Almería, Südostspanien. *Madrider Mitteilungen* 39: 35–48.

Stone, E., 1987, Nippur Neighborhoods. Chicago Oriental Institute, *Studies in Ancient Oriental Civilization* 44.

Stone, E., 1990, The Tell Abu Duwari Project, Iraq, 1987. *Journal of Field Archaeology* 17: 141–162.

Stone, E., and Zimansky, P., 1994, The Tell Abu Duwari Project, 1988–1990. *Journal of Field Archaeology* 21(4): 437–455.

Stone, E., and Zimansky, P., 1995, The Tapestry of Power in a Mesopotamian City. *Scientific American* 269(4): 118–123.

Stone, G., 1994, Agricultural Intensification and Perimetrics: Ethnoarchaeological Evidence from Nigeria. *Current Anthropology* 35: 317–324.

Strathern, A., 1969, Finance and Production: Two Strategies in New Guinea Highland Exchange Systems. *Oceania* 40: 42–67.

Strathern, A., 1978, "Finance and Production" Revisited: In Pursuit of a Comparison. *Research in Economic Anthropology* 1: 73–104.

Strathern, M., 1981, Self-Interest and the Social Good: Some Implications of Hagen Gender Imagery. In *Sexual Meanings: The Cultural Construction of Gender and Sexuality,* edited by S. Ortner and H. Whitehead, pp. 166–191. Cambridge University Press, Cambridge.

Stuart, D., 1988, The Rio Azul Cacao Pot: Epigraphic Observations on the Function of a Maya Ceramic Vessel. *Antiquity* 62: 153–157.

Stuart, D., 1993, Historical Inscriptions and the Maya Collapse. In *Lowland Maya Civilization in the Eighth Century AD*, edited by J. A. Sabloff and J. S. Henderson, pp. 321–354. Dumbarton Oaks, Washington, DC.

Stuart, D., 1999, "The Fire Enters His House": Architecture and Ritual in Classic Maya Texts. In *Function and Meaning in Classic Maya Architecture*, edited by S. Houston, pp. 373–425. Dunbarton Oask Research Library and Collection, Washington, DC.

Stuart, D., and Houston, S., 1994, *Classic Maya Place Names*. Dumbarton Oaks Research Library and Collection, Washington, DC.

Stuart, D. E., and Gauthier, R. P., 1984, *Prehistoric New Mexico: Background for Survey*. University of New Mexico Press, Albuquerque.

Stubbs, S., and Stallings, Jr., W. S., 1953, The Excavation of Pindi Pueblo, New Mexico. *Monographs of the School of American Research* 18, Santa Fe.

Sullivan, L. A., 1991, *Preclassic Domestic Architecture at Colha, Belize*. Master's thesis, University of Texas, Austin.

Sutton, P., 1988, *Dreaming: The Art of Aboriginal Australia*. George Braziller, New York.

Tedlock, D., 1985, *Popol Vuh: The Mayan Book of the Dawn of Life*. Simon & Schuster, New York.

Tello, J. C., 1956, *Arqueología del Valle Casma: Cultural Chavín, Santa o Huaylas, Yunda, y Sub-Chimu*. Publication Antropológica del Archivo "Julio C. Tello" de la Universidad Nacional Mayor de San Marcos, I. Universidad Nacional Mayor de San Marcos, Lima.

Teltser, P. A. (ed.), 1995, *Evolutionary Archaeology: Methodological Issues*. University of Arizona Press, Tucson.

Terral, J. F., 1996, Wild and Cultivated Olive (*Olea europaea* L.): A New Approach to an Old Problem Using Inorganic Analyses of Modern Wood and Archaeological Charcoal. *Review of Palaeobotany and Palynology* 91: 383–397.

Thomas, J., 1991, *Rethinking the Neolithic*. Cambridge University Press, Cambridge.

Thorpe, I. J. N., 1997, From Settlements to Monuments: Site Succession in Late Neolithic and Early Bronze Age Jutland, East Denmark. In *Semiotics of Landscape: Archaeology of Mind*, edited by G. Nash, pp. 71–79. BAR International Series, Volume 661, Oxford..

Tjumenev, A. I., 1969, The Working Personnel on the Estate of the Temple of Ba-U in Lagash During the Period of Lugalanda and Urukagina. In *Ancient Mesopotamia. Socio-Economic History*, edited by I. M. Diakonoff, pp. 136–169. Nauka, Moscow.

Topic, J., 1989, The Ostra Site: The Earliest Fortified Site in the New World? In *Cultures in Conflict: Current Archaeological Perspectives*, edited by D. C. Tkaczuk and B.C. Vivian, pp. 215–228. Proceedings of the Twentieth Annual Chacmool Conference. The Archaeological Association of the University of Calgary, Calgary.

Topic, J., and Topic, T., 1987, The Archaeological Investigation of Andean Militarism: Some Cautionary Observations. In *The Origins and Development of the Andean State*, edited by J. Haas, S. Pozorski and T. Pozorski, pp. 47–55. Cambridge University Press, Cambridge.

Treherne, P., 1995, The Warrior´s Beauty: The Masculine Body and Self-Identity In Bronze-Age Europe. *Journal of European Archaeology*, 3:1.

Trigger, B., 1990, Monumental Architecture: A Thermodynamic Explanation of Symbolic Behavior. *World Archaeology* 22: 119–132.

Trigger, B., 1998, *Sociocultural Evolution*. Blackwell, Oxford.

Turner, V., 1995, *The Ritual Process: Structure and Anti-Structure*. Aldine de Gruyter, Hawthorne, NY.

Tylor, E. B., 1871, *Primitive Culture: Researches into the Development of Mythology, Philosophy, Religion, Art, and Custom*. J. Murray, New York,

REFERENCES

Tylor, E. B., 1881, *Anthropology: An Introduction to the Study of Man and Civilization*. D. Appleton, New York.
Tymowski, M., 1991, Wolof Economy and Political Organization: The West African Coast in the Mid-Fifteenth Century. In *Early State Economics*, edited by H. J. M. Claessen and P. van de Valde, pp. 131–142. Political and Legal Anthropology Series, Volume 8, Transaction, New Brunswick.
Upham, S., ed., 1982, *Polities and Power*. Academic Press, New York.
Upham, S., 1990, *The Evolution of Political Systems*. Cambridge University Press, New York.
Valeri, V., 1985, *The Human Sacrifice: Ritual and Society in Ancient Hawaii*. Chicago University Press, Chicago.
Van De Mieroop, M., 1993, Sheep and Goat Herding According to the Old Babylonian Texts from Ur. *Bulletin on Sumerian Agriculture* 7: 161–182.
Van Zantwijk, R., 1985, *The Aztec Arrangement: The Social History of Pre-Spanish Mexico*. University of Oklahoma Press, Norman.
Vanderbroeck, Paul J. J., 1987, *Popular Leadership and Collective Behavior in the Late Roman Republic (ca. 80–50 BC)*. Dutch Monographs on Ancient History and Archaeology, edited by P. W. De Neeve and H. W. Pleket, Volume III. J. C. Gieben, Amsterdam.
Vandkilde, H., 1996, *From Stone to Bronze. The Metalwork of the Late Neolithic and Earliest Bronze Age in Denmark*. Jutland Archaeological Society Publications XXXII. Aarhus University Press, Aarhus.
Vernet, J-L., 1997, *L'Homme et la Forêt Méditerranéenne de la Préhistoire à Nos Jours*. Éditions Errance, Paris.
Vicent García, J. M., 1995, Early Social Complexity in Iberia: Some Theoretical Remarks. In *The Origins of Complex Societies in Late Prehistoric Iberia*, edited by K. T. Lillios, pp. 177–183. International Monographs in Prehistory, Ann Arbor.
Vilá Valentí, J., 1961, La Lucha contra la Sequía en el Sureste de España. *Estudios Geográficos* 22: 25–44.
Voigt, M., and Ellis, R., 1981, Excavations at Gritille, Turkey: 1981. *Paléorient* 7: 87–100.
Vradenburg, J. J., Benfer, Jr., R., and Sattenpiel, L., 1997, Evaluating Archaeological Hypotheses of Population Growth and Decline on the Central Coast of Peru. In *Integrating Archaeological Demography: Multidisciplinary Approaches to Prehistoric Populations*, edited by R. R. Paine, pp. 150–174. Center for Archaeological Investigations, Occasional Paper 24, Carbondale.
Waldren, W. H., Ensenyat, J. A., and Kennard, K. A. (eds.), 1995, *Ritual, Rites and Religion in Prehistory: III Deya Conference of Prehistory* (BAR International Series, Volume 611). Tempus Reparatum, Oxford.
Walters, K. R., 1996, Time and Paradigm in the Roman Republic. *Syllecta Classica* 7: 69–97.
Warren, A. H., 1969, Tonque: One Pueblo's Glaze pottery Industry Dominated Middle Rio Grande Commerce. *El Palacio* 76: 36–42.
Warren, A. H., 1979, The Glaze Paint Wares of the Upper Middle Rio Grande. In *Archaeological Investigations in Cochiti Reservoir* 4, edited by Jan V. Biella and Richard C. Chapman, pp. 169–175. Office of Contract Archaeology, University of New Mexico, Albuquerque.
Wason, P. K., 1994, *The Archaeology of Rank*. Cambridge University Press, Cambridge.
Watkins, F. M., 1968, State: The Concept. *International Encyclopedia of the Social Sciences* 15: 150–156.
Wattenmaker, P., 1987, Town and Village Economies in an Early State Society. *Paléorient* 13(2): 113–122.
Waylen, P. R., and Caviedes, D. N., 1986, El Niños and Annual Floods on the North Peruvian Littoral. *Journal of Hydrology* 89: 141–156.
Webb, M. C., 1975, The Flag Follows the Trade: An Essay on the Necessary Interaction of Military and Commercial Factors in State Formation. In *Ancient Civilizations and Trade*,

edited by J. A. Sabloff and C. C. Lamberg-Karlovsky, pp. 155–210. University of New Mexico Press, Albuquerque.

Weber, M., 1947, *The Theory of Social and Economic Organization*, edited by T. Parsons. Free Press, New York.

Webstein, D., 1999, Warfare and Status Rivalry: Lowland Maya and Polynesia Comparisons. In *Archiac States*, edited by G. M. Feinman and J. Marcus, pp. 311–351. School of American Research Press, Santa Fe.

Weigand, P. C., Harbottle, G., and Sayre, E. V., 1977, Turquoise Sources and Source Analysis: Mesoamerica and the Southwestern USA. In *Exchange Systems in Prehistory*, edited by T. K. Earle and J. E. Ericson, pp. 15–34. Academic Press, New York.

Wendorf, F., and Reed, E. K., 1955. An Alternative Reconstruction of Northern Rio Grande Prehistory. *El Palacio* 62: 131–173.

Westenhotz, A., 1984, The Sargonic Period. In *Circulation of Goods in Non-Palatial Context in the Ancient Near East*, edited by A. Archi, pp. 17–30. Edizioni dell'Ateneo, Rome.

Wetherington, R. K., 1968, *Excavations at Pot Creek Pueblo*. Fort Burgwin Research Center 6, Taos.

Whallon, R., 1979, *An Archaeological Survey of the Keban Reservoir Area of East Central Turkey*. University of Michigan, Museum of Anthropology Memoir no. 11, Ann Arbor.

Wheatley, M. J., 1994, *Leadership and the New Science: Learning about Organization from an Orderly Universe*. Berrett-Kohler Publishers Inc, San Francisco.

White, J. C., 1995, Incorporating Heterarchy into Theory on Socio-Political Development: The Case from Southeast Asia. *Heterarchy and the Analysis of Complex Societies*, edited by R. M. Ehrenreich, C. L. Crumley, and J. E. Levy, Archaeological Papers of the American Anthropological Association No. 6, pp. 101–123.

White, L., 1932, *The Pueblo of San Felipe*. Memoirs of the American Anthropological Association, No. 38. Menasha, WI.

White, L., 1935, *The Pueblo of Santo Domingo, New Mexico*. Memoirs of the American Anthropological Association, No. 43, Menasha.

White, L., 1942, *The Pueblo of Santa Ana*. Memoirs of the American Anthropological Association, No. 60, Menasha.

White, L., 1949, *The Science of Culture*. Grove Press, New York.

White, L., 1959, *The Evolution of Culture*. McGraw-Hill, New York.

White, L., 1962, *The Pueblo of Sia, New Mexico*. Bureau of American Ethnology Bulletin No. 184. Washington, DC.

Widmer, R. J., 1988, *The Evolution of the Calusa*. University of Alabama Press, Tuscaloosa.

Wilcox, D. R., 1991, Changing Contexts of Pueblo Adaptations, AD 1250–1600. In *Farmers, Hunters, and Colonists*, edited by K. A. Spielmann, pp. 128–154. University of Arizona Press, Tucson.

Wilkinson, T. J., 1990, *Town and Country in Southeastern Anatolia. Volume I: Settlement and Land Use at Kurban Höyük and Other Sites in the Lower Karababa Basin*. Oriental Institute of the University of Chicago, Chicago.

Wilkinson, T. J., 1994. The Structure and Dynamics of Dry Farming States in Upper Mesopotamia. *Current Anthropology* 35 (5): 483–520.

Wilkinson, T. J., and Tucker, D. J., 1995, Settlement Development in the North Jazira, Iraq. *Iraq Archaeological Reports* 3, British School of Archaeology in Iraq. Aris and Phillips, Warminster.

Willey, G., 1953, Prehistoric settlement pattern in the Viru Valley, Peru. *Bureau of American Ethnology Bulletin* 155:

Willey, G., 1962, Mesoamerica. In *Courses Toward Urban Life*, edited by R. J. Braidwood and G. R. Willey, pp. 84–101. Aldine, Chicago.

REFERENCES

Willroth, K. H., 1997, Prunkbeil oder Stosswaffe, Pfriem oder Tätowierstift, Tüllengerät oder Treibstachel? Anmerkungen zu einigen Metallobjekten der älteren nordischen Bronzezeit. In *Studia honoraria. Festschrift für Bernhard Hänsel*, edited by C. Becker et al. Internationale Archäologie 1. Verlaga Marie Leidorf GmbH. Espelkamp.

Wills, W. H., 1988, *Early Prehistoric Agriculture in the American Southwest*. School of American Research Press, Santa Fe.

Wilson, D. J., 1988, *Prehispanic Settlement Patterns in the Lower Santa Valley, Peru*. Smithsonian Institution Press, Washington, DC.

Wilson, P. J., 1988, *The Domestication of the Human Species*. Yale University Press, New Haven.

Winter, I., 1985, After the Battle is Over: The Stele of the Vultures and the Beginning of Historical Narrative in the Art of the Ancient Near East. In *Pictorial Narrative in Antiquity and the Middle Ages*, edited by H. Kessler and M. Simpson, pp. 11–32. National Gallery of Art, Studies in the History of Art, Volume 16, Washington, DC.

Winter, I., 1987, Women in Public: The Disk of Enheduanna, the Beginning of the Office of EN-Priestess and the Weight of Visual Evidence. In *La Femme dans le Proche Orient Antique*, edited by J. M. Durand, pp. 189–201. Editions Recherche sur les Civilisations, Paris.

Winterhalder, B., 1997, Gifts Given, Gifts Taken: The Behavioral Ecology of Nonmarket, Intragroup Exchange. *Journal of Archaeological Research* 5: 121–168.

Wistrand, E., 1979, *Caesar and Contemporary Roman Society*. Acta Regiae Societatis Scientiarum et Litterarum Gothoburgensis, Humaniora 15. Göteborg: Kungl. Vetenskaps-och Vitterhets-Samhället.

Woolley, L., 1934, *Ur Excavations, Volume II. The Royal Cemetery*. The British Museum and The University of Pennsylvania University Museum, London and Philadelphia.

Wright, H., 1969, The Administration of Rural Production in An Early Mesopotamian Town. *University of Michigan Museum of Anthropology Anthropological Paper No. 38*, Ann Arbor.

Wright, H., 1977, Recent Research on the Origins of the State. *Annual Review of Anthropology* 6: 379–397.

Wright, H., 1978, Toward an Explanation of the Origin of the State. In *Origins of the State*, edited by R. Cohen and E. Service, pp. 49–68. Institute for the Study of Human Issues, Philadelphia.

Wright, H., 1981, Appendix: The Southern Margins of Sumer. In *Heartland of Cities*, edited by R. M. Adams, pp. 295–345. University of Chicago Press, Chicago.

Wright, H., 1984. Prestate Political Formations. In *On the Evolution of Complex Societies. Essays in Honor of Harry Hoijer*, edited by W. Sanders, H. Wright, and R. M. Adams, pp. 41–77. Undena Press, Malibu.

Wright, H., and Johnson, G., 1975, Population, Exchange, and Early State Formation in Southwestern Iran. *American Anthropologist* 77: 267–289.

Wright, L. E., 1991, *Human Skeletal Remains and Preclassic Mortuary Practices from the 1989 Excavations at Operation 2031, Colha, Belize*. On file, Texas Archaeological Research Laboratory, The University of Texas at Austin.

Wright, R., 1996, Technology, Gender, and Class: Worlds of Difference in Ur III Mesopotamia. In *Gender and Archaeology*, edited by R. Wright, pp. 79–110. University of Pennsylvania Press, Philadelphia.

Yoffee, N., 1979, The Decline and rise of Mesopotamian Civilization: An Ethnoarchaeological Perspective on the Evolution of Social Complexity. *American Antiquity* 44(1): 5–35.

Yoffee, N., 1994. Memorandum to Murray Gell-Mann concerning the complications of complexity in the prehistoric Southwest. In *Understanding Complexity in the Prehistoric*

Southwest, edited by G. Gumerman and M. Gell-Mann, pp. 341–358. Addison-Wesley, Reading.

Yoffee, N., 1995, Political Economy in Early Mesopotamian States. *Annual Review of Anthropology* 24: 281–311.

Zaccagnini, C., 1983, Patterns of Mobility Among Ancient Near Eastern Craftsmen. *Journal of Near Eastern Studies* 42(4): 245–264.

Zeder, M., 1988, Understanding Urban Process Through the Study of Specialized Subsistence Economy in the Near East. *Journal of Anthropological Archaeology* 7: 1–55.

Zeder, M., 1991, *Feeding Cities: Specialized Animal Economy in the Ancient Near East.* Smithsonian Institution Press, Washington.

Zeder, M., 1994, Of Kings and Shepherds: Specialized Animal Economy in Ur III Mesopotamia. In *Chiefdoms and Early States in the Near East: The Organizational Dynamics of Complexity,* edited by G. Stein and M. Rothman, pp. 175–191. Prehistory Press, Madison.

Zilhão, J., 1993, The Spread of Agro-pastoral Economies Across Mediterranean Europe: A View From the Far West. *Journal of Mediterranean Archaeology* 6 (1): 5–63.

Index

Agency, 13, 15, 86, 214, 215, 236, 238
Aggregation, 15, 16, 38, 40, 48, 58, 153, 185
Alliance, 48, 56, 58, 161
Alto Salaverry, 188, 190-192, 197
Anagenesis, 20
Andean coast, 179, 186, 196, 197, 202, 203, 204
Andes, 12, 119, 150, 186, 187, 191, 202, 240
Argaric, 63-73, 75-81
Arroyo Hondo, 53
Autopoesis, 20, 21
Axis mundi, 141, 145, 163

Bands, 7, 15, 16, 24
Basin of Mexico, 142, 158
Belize, 130, 133, 136, 145
Bottom-up, 205, 236
Bourdieu, Pierre 109
Bronze Age,
    Spain, 59, 63-69, 72, 75, 77, 79-81,
    Europe, 85, 88, 92, 93, 95-100, 102-104, 110, 113-116, 124, 138
    Mesopotamia, 229, 237, 238
Burials, 107
    Spain, 66, 74, 77, 78, 80
    Europe, 85, 87, 88, 90, 92, 93, 96, 98-100, 110, 113, 115, 116
    Hawaii, 120
    Maya, 132, 133,
    Mesoamerica, 152, 158, 160, 164, 166, 171, 172

Calendar, 141, 168-170
Calendrics, 168
Casma Valley, 191, 193, 194, 198
Centralization, 11, 12, 17, 18, 36, 55, 58, 75, 83, 105, 126, 127, 147, 149, 150, 153, 154, 160, 161, 171, 172, 175, 178, 179, 181, 199, 201-203, 210, 215, 220, 235, 237-243
Centralized decision making, 235
Chang, K. C., 126, 128, 130, 134, 138-140, 144
Chaos, 21-23, 31-33, 201, 202
Chapman, Robert, 61, 63, 66-70, 72-75
Chiefdoms, 7, 9, 24, 36, 55, 56, 77, 84, 103, 106, 107, 110-113, 115-117, 119, 120, 122-124, 149, 152, 154, 156-158, 177, 194, 205, 215, 238-240, 242
Chiefly cap, 93
Chiefly priest, 93, 238
Chiefs, 12, 35, 83, 91-93, 95, 97, 99- 102, 112, 120, 122, 154, 159
Childe, V. Gordon, 5, 6, 65, 68, 76
China, 12, 126, 128, 140, 149, 240
Cities, 12, 16, 29, 30, 37, 111, 161, 170, 171, 208, 217-221, 224, 225, 227, 230, 231
Classic Maya, 126, 127, 134, 136, 139-145, 151, 152, 161, 163-166, 168, 170, 173, 174
Classic Period,
    Rio Grande, 37-39, 41, 48, 55,
    Maya, 128, 129-133, 135, 136, 139, 140, 142, 143, 145,

283

Classic Period (*cont.*)
  Mesoamerica, 153, 161, 164, 166-169
Colonialism, 19, 36, 40
Conflict, 26, 42, 51, 55, 58, 74, 76, 81, 138, 154, 183, 184, 196, 198, 200, 201, 203, 214, 220, 221
Conquest, 29, 116, 123, 161, 185, 199
Craft specialization, 17, 158
Crete, 93, 97, 98
Crowns, 161, 166
Cultural evolution, 3-11, 13, 16, 18, 31, 35, 105, 153, 160, 161, 236, 242

Darwinian, 10, 11, 20, 21
Defense, 53, 55, 58, 76, 113, 119, 184, 185, 196, 200
Denmark, 69, 84, 97, 113, 116, 124, 239
Dialectic, 22
Dualism, 91

Economic power, 180, 181, 185, 198, 203, 217
Egypt, 6, 12, 64, 171, 207
El Argar, 63, 72
El Mirador, 130, 135, 138, 145
Emblem glyphs, 135
Emulation, 86
England, 29, 110, 116, 117
Environment, 16, 57, 106-109, 111-113, 123, 124, 186, 187, 196, 199, 215, 226, 227, 239, 240
Evolution of power, 16, 18

Feasting, 136, 147, 159
Feinman, Gary, 9, 11, 16, 57, 74, 134, 149, 153, 155, 158, 172, 179, 215, 238-241
Formative period
  Maya, 129, 130, 132-136, 139, 143, 147, 148
  Mesoamerica, 158
Fortifications, 63, 66, 76, 102, 116, 184, 196, 199
Fried, Morton, 7-9, 154, 159, 167, 179
Fuzzy networks, 214

Galisteo, 39, 40, 42, 43, 47-51, 54, 57, 58
Gatas, 70, 72, 73, 79, 80
Glaze ware, 43, 45, 46
Government, 4, 17, 24, 28, 150, 156, 170, 213, 239, 242
Great Chain of Being, 19

Haas, Jonathan, 9, 11, 14, 17, 19, 39, 40, 51, 55, 68, 69, 74, 154, 175, 180, 185, 197, 198, 204
Harappa, 170, 173
Harris, Marvin, 5, 79
Hawaii, 84, 112, 123, 124, 239
Heiau, 121-123
Helms, Mary, 56, 86, 112, 127, 141, 159
Heterarchy, 24, 25, 28, 32, 56, 242, 243
Hierarchy, 19, 20, 24, 25, 28, 30, 32, 38, 48, 56, 63, 66, 68, 75, 76, 108, 110, 116, 120, 122, 130, 132, 134, 148, 152-154, 170, 179-181, 195, 198, 211, 239, 242
Hieroglyphs, 135, 141, 142, 239
Hill-fort, 116, 117
Hittite, 90, 97, 98
Hydraulic, 70, 72

Iconography, 98, 130, 139, 141, 143, 144, 149, 161, 167
Ideological power, 182
Impressed Ware, 61, 62, 73
India, 6, 12
Indus Valley, 170
Institutionalization of power, 110
Institutions, 5, 12, 61, 76, 77, 80, 81, 85-87, 97, 98, 103, 104, 106-108, 110-113, 122, 123, 177, 178, 183, 199, 205, 207, 210, 214, 216, 217-220, 222, 223-225, 227, 231, 237
Integration, 6, 7, 15, 16, 31, 50, 74, 105, 122, 159, 172, 179, 207, 215, 216, 220, 221, 225
Intensification, 15, 17, 50, 67, 68, 70, 71, 73, 74, 76, 78, 81, 110, 119, 188
Intermarriage, 161
Iron Age, 77, 80, 87, 93, 97, 98, 102, 103, 116
Irrigation, 7, 61, 67, 70-72, 74, 107, 112, 119, 121, 172, 180, 181, 186, 191, 192, 197, 198, 203, 241

Jacob, Francois, 13
Jemez, 38, 42, 45-48, 50, 51, 54, 57
Jutland, 88, 93, 113, 114

K'axob, 130, 132, 133
Kauffman, Stuart, 20-22, 24
King List, 210

INDEX

Kingship, 125-127, 130, 138-140, 142, 143, 145, 146, 148, 210, 211

Landscape, 36, 51, 84, 100, 101, 106-108, 110-116, 119, 120, 122, 124, 127, 195, 210, 216, 236, 240, 243
Legitimacy, 30, 127, 210, 215
Liminality, 28, 125
Lineages, 80, 81, 115, 170
Los Millares, 36, 63, 64, 68, 69, 71-73, 76, 78

Matrix organization, 27
Maya Long Count, 168
Mediterranean, 29, 31, 59, 61, 63-65, 81, 93, 98-100, 103
Megaliths, 62, 63, 80
Mesopotamia, 6, 12, 150, 171, 206, 207, 209, 210-214, 219, 220-223, 225, 229, 231, 241, 242
Metallurgy, 62-64, 66, 68-70
Military power, 183-185, 196, 198, 201, 203
Minoan, 30, 90, 93, 97, 98, 157
Moche Valley, 188, 190-192, 194-196, 198, 199
Monumental architecture, 130, 133, 135, 147, 150, 240
Monuments, 62, 63, 86, 106, 107, 110-112, 116, 119, 123, 124, 140, 141, 164, 171, 178, 182, 183, 185, 191, 193, 194-198
Morgan, Lewis Henry, 3, 4, 5, 8, 103
Mounds, 115, 116, 158, 192-194, 229, 240

Nakbe, 130, 132, 135, 139, 145
Neo-Darwinian, 10
Neolithic, 61-67, 72, 73, 87, 96, 110, 111, 113, 115, 126, 157, 237
New Guinea, 158
New Mexico, 36-38, 41, 57

Oaxaca, 158, 165

Palaces, 97, 99, 145, 164, 209, 213, 216, 217-225, 231
Palenque, 133, 134, 161, 163, 164
Paleobotanical, 70, 73
Permutation, 86
Peru, 84, 112, 116, 118, 124, 150, 186-188, 190, 193, 194, 196, 202, 239-242
  Late Preceramic, 178, 188-192, 195-197
  Initial Period, 178, 191
Polanyi, Karl, 179

Political authority, 126, 128, 130, 150, 167, 238
Political control, 102, 179, 180, 198
Political power, 43, 140, 147, 169, 180, 197, 203
Popol Vuh, 138, 139
Possehl, Gregory, 16, 171
Postprocessual, 109, 154
Processual, 80, 110
Pyramids, 110, 130, 132, 133, 135, 136, 145

Religion, 11, 12, 20, 29-31, 38, 48, 55, 56, 83, 96, 242
Renfrew, Colin, 65, 67, 75, 110, 112, 134, 156-158, 160
Risk, 74, 138, 146, 183, 200, 201, 203, 204, 227-229, 231
Rome, 29, 30, 139, 169, 170, 175
Ruling elite, 106, 150

Scandinavia, 83, 90, 95-98, 100, 101, 103
Selectionism, 15, 153
Sequential occupation, 39-41
Service, Elman, 7-9, 31, 154, 172, 179, 236
Shaman kings, 126
Shamanism, 125, 142
Shamanistic politics, 126
Site catchment, 67
Social hierarchies, 16, 17
Social stratification, 12, 70, 75, 77, 79, 214, 237
Social structure, 106
Southeast Spain, 59-64, 66, 67, 70, 72-80
  Copper Age, 62-65, 67, 69, 72, 73, 75, 76, 81, 237
Staple finance, 112, 116, 117, 119, 120, 124, 172, 180
States, 12, 17, 22, 24, 26-29, 31, 32, 36, 103, 106, 107, 120, 140, 149, 150, 236, 240, 242
  Spain, 77, 80,
  Mesoamerica, 152, 154, 156, 161, 167, 171,
  Peru, 177, 199, 203,
  Mesopotamia, 205, 207, 210, 211, 213, 215, 220, 222, 225, 231
Steward, Julian, 6, 7, 10

Temples, 30, 120, 133, 134, 167, 209, 210, 212, 213, 217, 218, 220, 222, 223, 225, 231
Temple state, 212

Thy, 94, 113-115
Tikal, 128-130, 136, 137, 139, 141, 163
Timing and tempo, 177, 178, 185, 203
Top-down, 205, 236
Trajectories, 12, 13, 16-18, 200, 202, 231
Transformation, 8, 11, 12, 14, 15, 18, 35, 55, 58, 103, 111, 115, 116, 126, 127, 144, 182, 238
Tribes, 4, 15, 17, 24, 55
Twin hat, 91
Twins, 138, 139
Tylor, Edward, 3, 4, 5, 8

Ubaid, 171-173
Urbanism, 211
Uruk, 208, 209, 211, 212, 223

Variation, 10-12, 14, 15, 56, 84, 102, 139, 141, 149, 156, 160, 161, 170, 173, 174, 178, 179, 186, 215

Wanax, 97, 98
Wanka, 116, 117, 119, 124
War chief, 93, 238
Warfare, 15. 83, 236, 238, 242
   Southwest United States, 38, 42, 56
   Spain, 73,
   Europe, 87, 93, 97, 100, 102, 103, 113,
   Maya, 135, 136, 138, 148,
   Peru, 184, 196, 198-200, 203,
   Mesopotamia, 211
Warrior aristocracies, 93, 96, 98-101, 103, 138
Warriors, 90, 98, 99, 102, 112, 120, 136-138, 148, 184
Wealth finance, 84, 124, 160
West Africa, 12
White, Leslie, 6, 7, 11, 16, 25-27
Willey, Gordon, 110, 151, 161, 190
Writing, 3, 59, 67, 107, 128, 140, 144, 152, 165, 182, 207-209, 211, 212, 220